System Efficiency by Renewable Electricity

Günther Brauner

System Efficiency by Renewable Electricity

Strategies for Efficient Energy Supply until 2050

 Springer

Günther Brauner
Institut für Energiesysteme
TU Wien, Wien, Austria

ISBN 978-3-658-35137-3   ISBN 978-3-658-35138-0   (eBook)
https://doi.org/10.1007/978-3-658-35138-0

Responsible Editor: Dr. Daniel Fröhlich
This Springer imprint is published by the registered company Springer Fachmedien Wiesbaden GmbH
part of Springer Nature.
The registered company address is: Abraham-Lincoln-Str. 46, 65189 Wiesbaden, Germany

Preface

The conversion of energy generation systems from fossil to predominantly renewable technology is concluded and necessary. The associated evolutionary process will require several decades, a high investment of capital and is very complex.

The past was characterized by the development of renewable energy sources, especially wind energy and photovoltaics, as well as the associated power electronics, protection and automation technology.

As the share of renewable energy generation increases, the ultimate goal of predominantly regenerative energy supply is getting closer and closer. Questions arise as to whether the renewable potential is sufficient, what the optimal generation scenarios and infrastructures of the future will look like, and whether all of this can be implemented in an environmentally friendly way and affordable for people.

The development of the energy turnaround so far makes it clear, that the initial studies that identified potential and developed overall supply scenarios from them are inadequate. The different characteristics of the renewable generation sources and the loads reduce the proportion of usable renewable energy. For example, a predominantly solar energy supply leads to a high oversupply in summer and to a lack of generation in winter. To shift the summer excess energy in the winter, large storage capacities would be necessary and would be used only inefficiently. The realization of major long-term storage capacities would be environmentally neither economically nor ecological feasible. Short-term centralized and decentralized storage capacities, on the other hand, are necessary in order to increase the utilization rate of renewable energy.

In this book, a more detailed analysis is carried out on the basis of quarter-hourly time series of the availability of wind energy and photovoltaics. The end use including electric mobility is also taken into account, with more detailed time series of load profiles for summer and winter and each for weekdays and weekends. This detailed analysis shows that photovoltaics in households can only be used economically to around 35% over the year. A utilization of 50% is possible together with local storage. The best inland generation mix consists of 75% energy from low-wind systems and 25% from building-integrated photovoltaics with local storage. This enables a renewable coverage rate of 80%.

A more detailed analysis of the renewable potentials from wind energy, photovoltaics, hydropower and biomass in Austria, Germany and Europe (EU-28) shows, that the economically and ecologically realizable renewable potentials are only sufficient for around 40 to 50% of today's final energy demand. Technological efficiency enhancement and changed sufficient user behavior are also required to enable a sustainable energy supply. By efficiency, energy will also be affordable for everyone. Energy-efficient buildings with heat pumps or storage heating, sustainable electromobility and sector coupling with power-to-heat, power-to-gas and power-to-mobility are the keys to technological success.

The potential analyzes also show that to achieve these goals, through photovoltaic roof and facade areas of the majority of existing buildings are needed and wind energy has to move closer to the settlements, as according to the current spatial planning rules. Here, conflicts can be anticipated, yet our idea of a livable settlement has been shaped by red roof tiles since Roman times. If the vast majority of houses have shiny blue photovoltaic modules and we no longer need roof tiles - will that find general acceptance? Can we develop new solar architectures, connecting the renewable production and life worth living?

There are also similar challenges with wind energy. In some countries, the minimum distance to settlements was set at 2000 m. However, the potential wind energy that this makes is no longer sufficient. The wind power installations must be allowed at least 1000 m from settlements. The familiar view of a vast country is changed by rotating wind turbines. Should we renew spatial planning rules to create tree belts that cause decoupling of wind generation from settlements?

These questions cannot be answered in this book, but arising from a more precise potential analyzes. They should be a stimulus for the architecture and the spatial and landscape planning.

All this shows, that the energy revolution has a lot of very complex tasks and cannot be planned by view of potentials and technology alone. Rather, it requires a long-term mediation process in which the advantages must be weighed against the disadvantages. New training content is required for fair mediation processes. Therefore, the Vienna University of Technology, together with the Diplomatic Academy of Austria offers a postgraduate training on *"Environmental Technology and International Affairs"*. On the one hand, environmental and energy technologies and on the other hand diplomatically skills are trained. The author of this book gives here lectures on *"Renewable Energy Systems"*. The book also contains content from this lecture.

Furthermore, lectures on *"Electrical Mobility"* and *"Solar Energy Systems"* at Danube University Krems are based on parts of this book.

Finally, I would like to thank the publisher Springer and Mr. Dr. Fröhlich for the excellent support and design of this book.

Vienna   Günther Brauner
25. February 2021

Contents

About the Author

Günther Brauner Technical University of Vienna Institute for Energy Systems and Electric Drives.

Study of communications engineering and doctorate in the field of high voltage engineering at the Technical University of Darmstadt. Afterwards 14 years at AEG Frankfurt in the field of network systems. There he was responsible for the development of the "program system for network planning tasks—PAN" and carried out system studies on network development, network dynamics, network control, blackout analysis and prevention. Since 1990 at the Technical University of Vienna in the field of energy systems with research work in the following areas: grid integration of renewable energy sources, in particular photovoltaics and wind energy, grid dynamics and grid control, master plans for transmission grids and metropolitan distribution grids, pumped storage, electromobility and decentralized regenerative energy systems. He was a member of the supervisory board of Verbund AG for 10 years and during this time gained experience in technical, economic and ecological system approaches and long-term corporate strategies.

Abstract of the Book: System Efficiency by Renewable Electricity

1

This chapter is a summary of the book and provides the key results. The background and detailed analyzes can be found in the corresponding chapters.

1.1 Energy Demand, Potential and Resources

Potentials and efficiency requirements

In the transition from fossil fuels to renewable energy supply, the technically, economically and environmentally friendly usable potentials represent the limits of feasibility. In this book, the possibilities and limitations are analyzed in long-term scenarios until 2050 by means of potential analyses and time-series simulations. From this, the necessities for efficient and sufficient end use of energy are derived.

The countries Austria and Germany are here compared. The European Union with its member states EU-27 and EU-28 are included in the comparison in terms of their total potential.

Table 1.1 shows the baseline situation in 2016. The electricity demand has a share of 20 to 24% in Austria, Germany and in the EU-28. Since the development of renewable energy supply is predominantly going in the direction of electricity generation with hydropower, wind energy and photovoltaics, a substitution of fossil energy with renewable electricity will be necessary in the future. For this, on the one hand, electricity generation from fossil fuels must be substituted by renewable energy. On the other hand, all previous fossil energy applications must be converted to renewable electricity as far as possible.

Table 1.1 shows that the sectors of final energy demand are comparatively very similar in all countries.

An analysis of the renewable potentials taking into account technical, economic and ecological boundary conditions in Table 1.2 shows that the potentials are limited. In almost all European countries the potential limits are around twice the

© Springer Fachmedien Wiesbaden GmbH, part of Springer Nature 2022
G. Brauner, *System Efficiency by Renewable Electricity*,
https://doi.org/10.1007/978-3-658-35138-0_1

Table 1.1 Energy supply 2016

	Austria	Germany	EU-28
Inhabitants (people PE) million	8.773	82522	511800
Area km^2	83,879	357,386	4,381,324
Population density PE/km^2	105	231	117
Total final energy demand TWh/a	327	2517	12,883
Renewable energy of the final energy demand	33.5%	14.8%	17.0%
Final energy demand per PE in MWh/a	37.30	30.5	25.2
Total electricity demand TWh/a	68	611	3070
Electricity requirement per PE in MWh/a	7.8	7.4	6.0
Electricity / final energy in % 2016	20.8%	24.3%	23.8%
Final energy demand shares 2016			
Household	25%	25%	25.4%
Traffic	33%	31%	33.1%
INDUSTRY	26%	29%	25.3%
Business, Trade, Services (BTS)	16%	15%	16.2%

Tab. 1.2 Potentials of renewable electricity by 2050

TWh/a	Austria		Germany		EU-28	
	2016	2050	2016	2050	2016	2050
Hydropower	39.3	42	20.6	22	340	500
Onshore wind energy	5.2	20	66.3	400	237	2000
Offshore wind energy	–	–	12.3	200	47	700
Photovoltaics	0.5	30	38.1	250	102	1500
Biomass	2.5	20	50.8	60	169	300
Geothermal energy	0	0	0.16	20		60
Renewable Electricity (RE)	**47.5**	**112**	**188.2**	**952**	**895**	**5060**
End electricity demand (EE)	68	140	611	1200	3070	6200
% EE of end electricity demand (RE/EE)	70%	80%	31%	79%	29.2%	82%

electricity demand of 2016. This means, that compared to the total final energy demand of 2016, around 50% has to be saved.

Figure 1.1 shows a conceivable energy strategy for the future. The current total energy demand is reduced by 50% through measures for efficiency and sufficiency and the usable renewable share is increased to 40% of the current final energy demand by expanding the regenerative energy sources according to technical and economic criteria and in compliance with applicable environmental standards.

Fig. 1.1 Change in energy demand in % in Europe by 2050

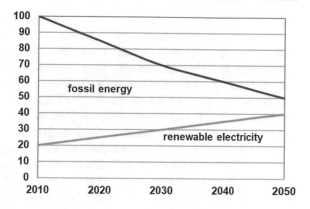

There is still a gap of 10% of today's demand between the usable regenerative potential and the demand. This results from the different temporal course of supply and demand. In particular, longer periods of regenerative overproduction can be followed by longer periods of underproduction. In the case of solar energy conversion, a summer–winter shift may be necessary, and wind energy has a higher supply in winter than in large-area high pressure areas in summer.

In relation to the future reduced final energy demand in 2050, this means that here only 80% can be produced from renewable sources (last line of Table 1.2). Closing this gap would be possible with long-term storage capacities. Studies in Austria have shown that pumped storage capacities would have to be increased by a factor of 150 for this. However, there are not that many valleys for flooding in Austria. These storage facilities would also be uneconomical, since they would only be filled about twice a year and would therefore have very high fixed costs. Ecology and economy only allow short-term storage capacities. From today's perspective, environmentally friendly reserve power plants must therefore continue to be available. These can be decentralized small power plants with fuel cells and hydrogen as energy sources, gas engines in cogeneration plants or building heating systems with natural gas or biogas, or central combined cycle gas-and-steam power plants of large power with district heating and with natural gas or biogas as fuel.

The reserve power plants would only be used for around 1000 to 1500 full load hours per year and would therefore have higher fixed costs. Because of the short hours of use and the resulting low share on total generation, this has only a minor impact on electricity prices.

As can be shown, a full renewable supply in 2050 is technically and economically hardly possible. Reserve power plants are therefore necessary and their output power should roughly correspond to the peak load of the grid. This enables a high level of security of supply, since longer periods without sufficient supply can then be safely bridged. This makes it clear, that a systemic approach, taking into account potential, costs, environmental impact and security of supply, leads to compromise solutions. In contrast, the ideology of a purely renewable energy supply would, from today's perspective, lead to uneconomic systems with high environmental impacts caused by long-term storage capacities. In the long term, a full regenerative supply is definitely possible. This requires new storage technologies, sector coupling and changed user behavior, which will have to develop over the next few decades and is hardly predictable.

The reduction potential of the energy demand by 2050 is expected to 70% by technologies for efficiency and by 30% by altered sufficiency-oriented user behavior. Technological efficiency improvements in end use demand are possible by switching mobility from the combustion engine to the electric drive. In the area of heating, by efficiency measures on buildings by thermal insulation and by introduction of electric heat pumps instead of oil or gas heating, the energy demand of buildings can be lowered. The voluntary limitation of the living space per person according to demand represents a measure of sufficiency. The industrial energy demand can be converted from fossil energy to efficient electricity applications, biomass derivatives and sector coupling to predominantly renewable energy supply.

Wind energy
The onshore wind turbines have in the long term to be developed technologically toward low wind turbines in order to need lower transmission capacities in the grid. The grid expansion, which is little accepted by the population, represents a major obstacle to the adequate development of the grid according to the expansion of renewable energy generation. Increased decentralization can partially mitigate this. However, central infrastructures for wide-area interconnection of renewable generation are still necessary.

The regional development planning presents another obstacle to the development of wind energy. Minimum distances from 2000 m to settlements limit the expansion possibilities. It would be necessary to reduce the minimum distances to about 1000 m.

Wind energy in inland Europe can be increased by a factor of nine and will account for 50 % of renewable electricity in 2050. Offshore wind energy will achieve a share of 15 % and building-integrated photovoltaics 30 %.

Photovoltaics
The ecological potential of photovoltaics is limited by the usable roof and facade areas of the buildings. In order to achieve the expansion targets, the majority of these areas are required for PV installations. Subsidized feed-in tariffs for PV installations are in the long term no more attractive, due to regular tariff reductions. In future PV systems are owned by the local residents or by energy service provider. As grid tariffs for electrical energy will rise in the future, local PV system will become economical. PV Contracting represents a new interesting business sector for utilities and investors. Since, due to the different characteristics of regenerative generation and energy demand, only about 50% of the building integrated PV can be used locally. The PV retail market with aggregators and

micro-balance groups will form a new segment of growth. Together with low wind systems and PV storage systems, coverage rates of the household load of up to 80% are possible.

Hydropower
In the field of energy supply, hydropower has historically been of great importance. The potential has largely been expanded in Germany. In Austria there is a residual potential of 15% that is worth expanding and in Europe this amounts to 40%.

Run-of-river power plants, both small and large, will continue to have an annual growth of 0.9% in Europe until 2050. Development is possible, especially in countries with still untapped potential.

The expansion of the pumped storage power plants is necessary for the short-term storage of renewable surplus energy and for the provision of grid control and balancing energy. In the period of predominantly fossil energy supply, pumped storage plants were designed to receive surplus night-time electricity from thermal power plants and to deliver it as peak load at noon the following day. The renewable energy supply requires larger storage capacities, in particular greater pumping capacities. Renewable generation from wind energy can continuously generate high surplus power over several days. Pumped storages must therefore be expanded from day to week storage.

Biomass
Significant shares of the heat supply for buildings and industry can be provided by biomass. A share of electricity generation of 6 to 10% is also possible via combined heat and power (CHP). Due to the limited expansion potential of wind energy and photovoltaics, biomass will be of increasing importance until 2050.

1.2   Energy Economy of Efficiency and Sufficiency

With the generation scenarios for 2050 in Table 1.2 and the expected developments in the levelized costs of electricity from renewable generation and the backup supply as well as the necessary grid expansion, the electricity prices including grid usage fees and taxes can be calculated up to the year 2050. Figure 1.2 shows that the electricity prices will almost double by the 2050. However, the electricity production costs from wind energy and photovoltaics will decrease significantly. This opens up a gap between grid tariffs and self-generation costs. PV systems without accumulators (PV) are already economical today. PV systems owned by the house residents with accumulators (PV & ACC POM), which can be marketed to the residents according to a public owner model (POM), will already be economical from 2020. PV systems with accumulators allow only a usage of about 50% of renewable generation due to the different characteristics of production and demand. But they will be economical in form of a public owner model from the year 2030.

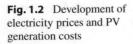

Fig. 1.2 Development of electricity prices and PV generation costs

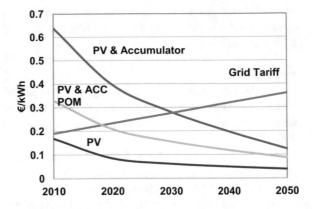

PV contracting represents in the future an attractive business field for energy service providers and investors. Contracting will also replace the model of subsidized feed-in tariffs: If everything is renewable, not everything can be subsidized anymore.

The higher electricity prices create incentives for efficient and sufficient end use. End users, who live in efficient buildings with efficient end-use equipment and an electric vehicle instead of a vehicle with an integrated combustion engine, have already in 2016 lower annual energy costs than the normal consumer. It is assumed that the depreciation on efficiency measures will be comparable in the future. Efficiency therefore creates affordability for everyone, and efficiency contracting is also a business area with strong growth potential.

Increasing urbanization supports efficiency and sufficiency. Compared to rural settlements with average values of 45 m² living space per inhabitant, this value drops to 38 m² in large cities. In rural regions, over 80% of the buildings are single or two-family houses with increased heating requirements. In large cities, multi-family houses with less outdoor surface per residential unit predominate. Vehicle density is also falling from 550 vehicles per 1000 inhabitants in rural regions to around 220 in large cities with attractive local public transport.

1.3   Storage Technologies

The renewable energy supply with its fluctuating characteristics requires storage technologies to adapt to consumption. Local house and vehicle batteries as well as pumped storage have to be developed adequately in terms of their capacities.

Stationary accumulators are required for balancing PV systems. Wind energy requires pumped storage systems. As Fig. 1.3 shows, pumped storages are in relation to accumulators in their specific capacity costs in future competitive, when moving their capacity form day to week storage, which means in the range of 20 to 150 h. The economic efficiency of the storage systems is determined by their

Fig. 1.3 Comparison of specific capacity costs of pumped storage and accumulators

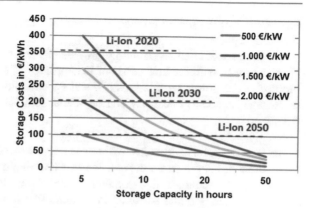

annual full load hours. Overcapacities are to be avoided both for decentralized accumulators and central pumped storage facilities.

1.4   Distribution and Transmission Grid

The development of grids for large transmission powers of renewable energy and for more spacious transport, especially for the coastal Onshore- and Offshore-wind energy, is an essential condition for the success of the energy transition.

For technical and economic reasons, high and extra high voltage grids should continue to be expanded with overhead lines wherever possible. The power of the wind turbines increases with the wind speed. The cooling of the overhead line conductors improves also strongly with increasing wind speed. At the nominal wind speed of 12 m/s, at which the wind turbines reach their rated power, over-head lines achieve roughly three times their rated transmission capacity. Cables in the 380 kV grid show no such behavior. Due to their high capacitive power demand, they are for high and extra-high voltage transmission grids not suitable. They also have very long repair times. Their investment costs are 5 to 8 times higher than those of overhead lines. Thus, they do not meet the technical and eco-nomic requirements for an efficient transmission grid infrastructure.

In the medium and low voltage range, on the other hand, cables are very well suited for network expansion. In regions with agricultural soil, cables can be plowed in cost efficient and allow sufficient large distances in distribution grids.

1.5   Decentralized Energy Supply

In future, the centrally controlled Smart Grid will be replaced by the decentral-ized self-organized Micro Grid. By developing decentral energy cells as Micro Grid with local renewable energy generation corresponding to the local demand, the necessary extension in the superimposed networks can be reduced to a feasible level (Fig. 1.4). Increased horizontal use of the distribution grids reduces the need for vertical use.

By time series analysis of energy provided by solar radiation or wind speed and by comparison with the time series of load profiles of households, the most favorable design for power installations of local PV modules and wind turbines as well as storage capacities necessary can be found. The following design criteria were found:

- In the case of PV systems without storage, an installation of PV power up to three times the average peak load of households in community residential buildings is economically expedient. Due to the different average characteristics of generation and consumption, higher PV outputs can hardly be used and flow into the grid. Since many—no longer subsidized—PV systems with such behavior will feed in, oversupply at marginal costs near to zero will be the result and the export of PV energy will in the future be uneconomical.
- PV systems without storage and with an economical size of the output of the PV systems according to twice the household peak load can only use around 35% of the annual PV generation locally. An increase to 50% is possible with local storage capacities. Storages having a capacity of up to three full load hours of the PV system are suitable. Larger storage capacities increase the share of usable PV only marginal and are therefore uneconomical.
- Onshore low wind turbines represent in future an appropriate renewable generation technology. The best production scenario is a combination of 75% of the annual energy demand by low-wind turbines and of 25% from building-integrated PV systems with local storages. This enables a high average coverage rate of the household load of 80% from renewable energy. With low-wind turbines alone, a coverage rate of only 70% is possible. Photovoltaic with local

Fig. 1.4 Horizontal grid usage with decentralized energy supply

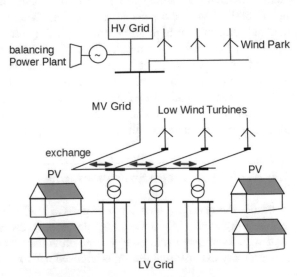

storage allows only 50%. This illustrates that not only the annual generation of the renewable sources is decisive for assessing the coverage rate, but that the characteristics of the generation and the demand are also important.

1.6   Buildings and Heating Demand

The thermal refurbishment of buildings to reduce the heating energy demand is necessary because of the limited renewable potential and the future higher energy costs.

The replacement of heating systems with fossil fuels by electric heat pumps results in very high saving potentials. Bivalent heat pumps can also be used for room cooling in summer. Electric storage heaters are also interesting in highly thermal insulated houses. They make it possible to store renewable surplus energy as heat and are useful for balancing energy grids. In energy-efficient buildings, heat pumps and storage heaters can be used economically even with higher network tariffs.

1.7   Sustainable Mobility

Motorized individual transport accounts for 65% of the total energy demand of the transport sector. The energy demand of the transport sector on the basis of mineral oil corresponds approximately to the future available renewable potential for electricity generation in Europe. Since 94% of the transport uses today fossil fuels, renewable concepts can simultaneously minimize energy demand and emissions. Figure 1.5 shows the possible reductions in the specific energy demand in the transport sector through the introduction of the electric drives (red) instead of the internal combustion engines (blue).

Vehicles with electric drive and accumulator can be used in local and commuter traffic. In long-distance traffic, fast charging stations are required to enable

Fig. 1.5 Efficiency potential when converting from combustion engine to electric drive

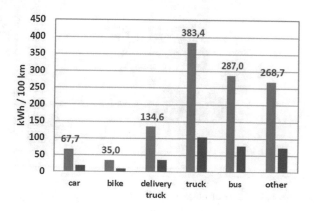

adequate charging times. For the heavy-duty transport sector, charging power of 500 kW would be required in accordance with the permitted driving times of 4.5 h and the subsequent minimum rest period of 45 min. An alternative here is the drive with hydrogen, fuel cell and electric motor. Short refueling times are possible here. However, the efficiency of production of hydrogen by electrolysis from renewable electricity together with the efficiency of the fuel cell, only allow overall efficiencies of around 30%. In comparison, the efficiency for charging vehicle batteries is around 80%. However, since heavy-duty transport accounts for less than 20% of the energy demand of the transport sector, hydrogen technology is justifiable from the point of view of renewable potential.

1.8   Sector Coupling

Through sector coupling, regenerative surplus energy can be used efficiently. The following couplings are promising:

- **Power-to-Heat**. In this case, renewable electricity can be converted into heat with a high degree of efficiency by electrical heating using heating rods or heat pumps. On a large scale, electrode boilers with outputs of up to 90 MW can provide process heat for industry or heating for district heating in a cost-effective and rapidly controllable manner. Figure 1.6 shows the use of an electrode boiler in a combined heat and power plant to use renewable surplus energy.
- **Power-to-Gas**. Electrolysis can produce hydrogen with an efficiency of 60%. Hydrogen can be fed directly into the existing natural gas pipeline in small quantities of a few percent. Methane can be produced through synthesis with carbon dioxide, which is very suitable and unlimited usable for storage in

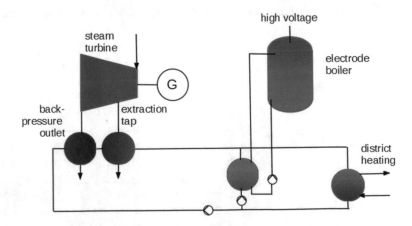

Fig. 1.6 Use of an electrode boiler in a combined heat and power plant

natural gas reservoirs. Hydrogen or methane can be used in fuel cell power plants for CHP or in electric vehicles with fuel cells.

- **Power-to-Liquid.** This includes various processes for the production of liquid hydrocarbons from hydrogen or synthetic methane using the Fischer–Tropsch process.

Through sector coupling, heavy industry can also be partially supplied with renewable surplus energy and thus its fossil energy requirements can be reduced.

Because of the limited potential of renewable energy, sector couplings in particular are to be promoted, which have a high level of efficiency in end use.

In addition to renewable electricity, biomass is also suitable for the production of gaseous or liquid fuels.

Energy Demand, Potential and Resources

2

2.1 Development of Energy Demand

The energy demand for Austria, Germany and the European Union of 28 countries (EU-28) is compared in the following. Table 2.1 shows the key figures for these countries in 2016. The share of electricity in the net final energy demand is in the order of 20 to 24%. In the lower part of the table, the distribution of the total final energy demand is shown in the individual sectors. It turns out, that all European countries have similar demand structures on average.

The final energy demand doubled in the period from 1970 to 2017. The demand for electricity has grown similarly to the total final energy demand.

The shares of the sectors of final energy demand are similar in Austria, Germany and the EU-28. Figure 2.1 shows the sectors in Germany in 2016 [EEZ 2016], each with the absolute share of electricity including fossil energy. The transport sector obtains 94% of its energy demand from fossil fuels. The sectors of industry, household and business, trade and services (BTS) are predominantly supplied with fossil fuels. Electricity generation itself has 2020 a share of renewable energy (RE) from wind, solar, hydropower and biomass of 40%. The total electricity generation is equivalent to 24% of final energy demand. In 2016 about 60% of the electricity generation plants were using fossil fuels.

Figure 2.2 shows the development of the relative share of electricity on total energy demand in Austria since 1970. The total energy demand has almost doubled in this period. Before 1970, when the expansion of large-scale hydropower began, the share of electricity was 13%. After that it stayed constant at 20%.

The constant percentage of electricity of 20 % indicates that the sector coupling between the generation of electricity and the sectors of end use as there are transport, space and process heat in the last 50 years has hardly developed.

© Springer Fachmedien Wiesbaden GmbH, part of Springer Nature 2022
G. Brauner, *System Efficiency by Renewable Electricity*,
https://doi.org/10.1007/978-3-658-35138-0_2

Table 2.1 Characteristic values in Europe 2016

	Austria	Germany	EU-28
Inhabitants (PE) million	8.773	82522	511800
Area km²	83,879	357,386	4,381,324
Population density PE/km²	105	231	117
Final energy demand TWh/a	327	2,517	12,883
Renewable energy of the final energy demand	33.5%	14.8%	17.0%
Final energy demand per PE in MWh / a	37.30	30.5	25.2
Electricity demand TWh/a	68	611	3070
Electricity demand per PE in MWh/a	7.8	7.4	6.0
Electricity/final energy % 2016	20.8%	24.3%	23.8%
Final energy demand shares 2016			
Household	25%	25%	25.4%
Traffic	33%	31%	33.1%
Industry	26%	29%	25.3%
Business, Trade, Services, BTS	16%	15%	16.2%

Fig. 2.1 Share of the sectors in the final energy demand and division into electrical and fossil energy [EEZ 2016]

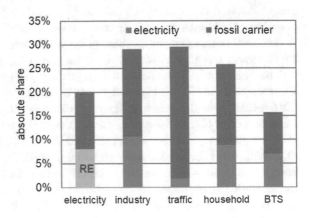

The future of energy supply lies in the sustainable development of generation. Hydropower, wind energy and photovoltaics are the regenerative energy sources that will experience strong growth in the coming decades, especially in the case of wind and PV. This requires stronger sector coupling. Vehicles must be converted from the internal combustion engine powered by mineral oil products to the electric vehicle powered by renewable electricity. This results in a reduction of the final energy demand by 70% with simultaneous emission-free driving.

The space heating and industrial process heating sectors also have to be converted to predominantly renewable electricity. In the case of space heating, the

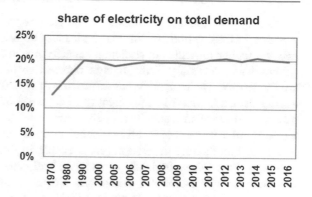

Fig. 2.2 Share of electricity in total final energy demand in Austria [Östat 2016]

main part can be achieved through the use of heat pumps. In the field of industrial high-temperature process heat, this is either direct electrical heat or the use of hydrogen and the synthesis products obtained from it, generated by electrolysis from regenerative electricity or by pyrolysis from biomass. Through the sector coupling, the share of electricity is continuously increasing, while the final energy demand steadily decreases. Figure 2.3 shows such a scenario up to the year 2050: "Renewable electricity applications substitute fossil energy applications".

As will be shown, it is hardly possible due to the limited potential of renewable energy, to cover the final energy demand of today fully by regenerative energy. Due to technical, economic and environmental potential limits only about a doubling of the today's electricity generation is possible, that means about 40% of the total energy demand 2016. As a result, on the one hand, today's fossil and nuclear electricity generation must be replaced and, on the other hand, the fossil final energy use must be converted to renewable electricity through sector coupling to the transport sector and the heating sector.

As Fig. 2.3 shows, energy supply predominantly from renewable electricity in particular from wind energy and photovoltaic is possible. Because of their

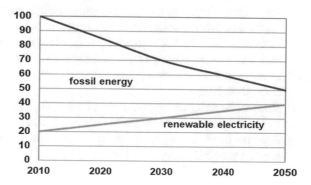

Fig. 2.3 Sustainable energy supply by 2050

fluctuating generation characteristic, a complete regenerative conversion is hardly possible as long periods of lull and dark must be bridged. Since long-term storage are neither to realize in sufficient capacity nor in an economic way, a backup generation is necessary. From today's perspective thermal generation plants are still necessary. Their capacity has to provide about 20% of the future annual end energy demand, which is equivalent to about 10% of the end use of today. The power of these plants should roughly correspond to the peak load of the network in order to be able to bridge longer periods without sufficient regenerative potential. The thermal generation plants enable in future a high level of security of supply. Small decentralized systems with fuel cells and hydrogen from surplus electricity or with motors for natural gas, biogas or liquid biofuels can be used as system types. Central thermal power plants are also possible as highly efficient combined cycle gas and steam power plants with district heating and with biogas or natural gas as fuel.

2.2   Development Goals of Renewable Energy Supply

The development of the sustainable energy supply is in most countries around the world a significant element of the future strategy. Since the energy transition is an evolutionary process with many influencing factors, the development goals are fuzzy and framework conditions change accordingly frequently. The following development goals can be considered long-term valid (REEES):

- **Renewable**: Fossil energy is replaced by sustainably generated electricity, since fossil energy is limits in its resources and the use is harmful to the climate.
- **Efficient**: The final energy demand is adjusted to the technically, economically and environmentally compatible available renewable potential through efficiency in the end use of energy.
- **Economical**: The renewable energy supply must be developed according to economic criteria. Subsidies are being replaced by temporary start-up financing in research, development and implementation. If the energy supply is predominantly regenerative, not everything can be subsidized. An economical regenerative energy supply enables an affordable energy supply for everyone.
- **Environmentally friendly**: The energy converters and the associated infrastructure must be implemented in an environmentally friendly manner. This also requires an increase in efficiency and a reduction in demand, so that the adverse effects on living environment, nature and landscape remain within acceptable limits.
- **Secure**: Security of supply, also in renewable energy generation, must continue to meet a high standard. Switching to renewable energy reduces the dependence on fossil energy supplies. Decentralized renewable energy cells with the highest possible share of self-generation can increase local security of supply.

Centralized and decentralized storage facilities can improve security of supply even with fluctuating generation. Fossil back-up supply can safely bridge longer renewable generation gaps.

Figure 2.4 shows the strategy to achieve a predominantly renewable energy supply. The previous electricity generation, accounting for about 20% of final energy demand, must be converted to renewable generation to achieve the new targets. If the potential limits are observed, electricity generation can be roughly doubled. To convert all sectors of final energy use to renewable energy, the potential limits must be exhausted.

This means further that the final energy demand, compared to the present level of today, has to be reduced by about 60%. An initial estimate shows, that 70% of this reduction can be achieved through efficiency technology, for example by replacing the combustion engine with an electric drive and the oil or gas heating with an electrically powered heat pump.

The remaining 30% of demand reduction are achieved by altered user behavior, for example through demand reduction by sufficiency, by reduction of the living space according to the actual demand, or by using a smaller electric vehicle instead of a large SUV on well-developed roads.

Figure 2.5 shows the required change in electricity generation and final energy demand of the sectors up to 2050, based on the demand in 2016. The electricity is mainly regeneratively (RE) generated by development of the limited residual potential of hydro power and by intensive expansion of wind energy and photovoltaic and partial use of biomass. The individual end-use sectors have experienced an increase in efficiency and only need 40% of the energy demand from 2016. Under these conditions, a predominantly renewable energy supply is possible. However, balancing energy is still required from thermal power plants that use biofuels and, in some cases, still small amounts of fossil fuels, in order to bridge longer-term periods with insufficient renewable generation. The balancing energy

Fig. 2.4 Limits of potential and efficiency targets (base 2015, target 2050)

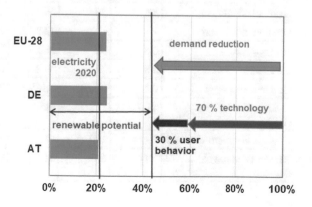

Fig. 2.5 Electricity generation and sector shares in the final energy demand in 2050

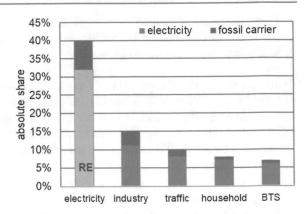

demand is around 20% of the final energy requirement by 2050, corresponding to 10% of the demand in 2016.

Table 2.2 shows the status of regenerative energy supply for Austria, Germany and the EU-28 in 2016. The potentials for expansion in 2050 are fuzzy, as they depend on the future framework conditions and the future acceptance of the population for the expansion of wind farms and PV plants, as well for the necessary expansion of the transmission grid and the storage capacities. Furthermore, high capital expenditures are required, which are not equally possible in all countries. In this scenario, the share of renewable electricity will reach around 80% of the final electricity demand in 2050.

Table 2.2 Renewable electricity in 2016 and potential up to 2050

TWh/a	Austria		Germany		EU-28	
	2016	2050	2016	2050	2016	2050
Hydropower	39.3	42	20.6	22	340	500
Onshore wind energy	5.2	20	66.3	400	237	2000
Offshore wind energy	–	–	12.3	200	47	700
Photovoltaics	0.5	30	38.1	250	102	1500
Biomass	2.5	20	50.8	60	169	300
Geothermal energy	0	0	0.16	20		60
Renewable electricity (RE)	**47.5**	**112**	**188.2**	**952**	**895**	**5060**
End-use electricity demand	68	140	611	1200	3070	6200
RE % of end-use electricity demand	70%	80%	31%	79%	29.2%	82%

2.3   Potential of Renewable Energy

2.3.1   Limits of Potential

The potentials of renewable energy are defined by several limit criteria. The smallest potential of it determines the possible expansion.

Theoretical potential: Theoretical physically useable renewable potential.

Technical potential: Available potential taking into account the efficiency of the generation facilities and other environmental framework conditions e.g., restricted zones such as approach corridors of airports. The efficiencies of the systems are assumed to correspond to the state of the art.

Economic Potential: Determined by the generation cost of renewable energy, resulting from the depreciation of the installed systems and other cost components for grid integration, taxes, insurance and levies of all kinds.

Ecological potential: Determined by the impact on the environment. In the case of biomass, the annual renewed potential is viewed as the limit for the possible harvest. In the cultivation of energy crops the usable area is determined by the competition with the further requisite of agricultural food production in the existing limited farmland.

In the case of sustainable electricity generation, the possible expansion results from framework directives such as the EU directive *"Framework in the field of water policy 2000/60/EU"*. Also, specifications for spatial planning for example the minimum distances to settlements for the expansion of wind parks fall into this category. The possible installation of PV systems on buildings can be determined by building regulations. Other environmentally relevant criteria can be the designation of "bird sanctuaries", nature reserves or "river sanctuaries" that prevent the installation of renewable generation systems. The ecological potential increasingly determines the possible expansion of regenerative energy.

The expansion of renewable energy is generally associated with extensive approval procedures. In an environmental impact assessment, a balance is struck between the public interest in a clean, regenerative energy supply and the interest in untouched nature or in acceptable living conditions for people in their settlements or in the surrounding landscape.

Due to their large number and widespread distribution, renewable energy plants have a much greater spatial impact than thermal power plants. Spatial planning is therefore becoming increasingly important in the approval process.

Realizable Potential: This is a small part of the potentials, that results as the intersection of the technical, economic and ecological potential and further requires the acceptance in the affected region.

The criteria for the potential limits are to be examined for wind energy, photovoltaics, hydropower and biomass.

2.3.2   Wind Energy

Spatial planning and potential in Austria

In the early days of wind energy, there were no specifications for the installation of wind turbines. There were individual permits. This led to very different assessments. It was recognized very early on that general rules in accordance with coordinated spatial planning could simplify procedures and increase investment security. As one of the first provinces in Austria, Burgenland defined in the year 2002 a binding regional spatial planning concept for the construction of wind turbines [TUW 2014]. Zones for wind energy use were designed in which several criteria had already been met, such as distances to settlements of 1000 m, access to electricity grid, avoidance of nature conservation areas, distances to existing wind turbines, excluding any flight paths and including the requirements of radar systems. All other areas were declared unsuitable for wind power plants and thus exclusion zones. However, the municipalities had the freedom to approve wind farms outside the designated zones within their municipal territory boundaries.

The province Lower Austria developed a similar spatial planning concept. In order to avoid a proliferation of wind turbines with low full-load hours, turbines may only be erected in areas that have at least an average annual power density of 220 Watts/m^2 at a hub height of 130 m above ground. The minimum distance to residential areas in the local community is 1200 m [TUW 2014]. If the residential area belongs to a neighboring municipality, a minimum distance of 2000 m must be observed. In the other federal provinces of Austria, the minimum distances to residential areas are uniformly 800 m.

About the realizable wind energy potential in Austria, several studies have been published with different results. New studies suggest a realizable potential of 6 GW and yields of 12 GWh/a. For this, 1.2% of the area of Austria is required [S4MG 2011]. A realizable potential of 6.65 GW with yields of 17.7 TWh/a is expected by 2030 [EW 2014]. In Table 2.2, a realizable potential of 8 GW with yields of 20 TWh/a at 2400 average full load hours is assumed by 2020. The types of turbines required for this are low wind turbines, which replace the previous wind energy technologies in the case of renewal or new construction. Wind energy will then in Austria have a share of 18% of electricity generation in 2050. However, hydropower remains the dominant renewable generation technology in Austria.

Spatial planning and potential in Germany

In Germany, spatial planning and the designation of zones for the use of wind energy have become very important. In 2011, by the German Wind Energy Association the potential of onshore wind energy was examined in detail [BWE 2011].

The theoretically usable areas of grassland, forest and usable protected areas correspond to 22% of the area of Germany. The technically usable potential is determined for 3 MW wind turbines. For areas with low wind potential these are designed as low wind turbines, with a hub height of 150 m and a specific rotor area of 3.5 m²/kW. In the other areas, 3 MW strong wind converters with a hub height of 100 m and a specific rotor area of 2.6 m²/kW are used.

The theoretical potential in Germany is 1500 GW (22% of the area), the technical potential on areas without restrictions is 722 GW (8% of the area). The realizable potential at a land use of 2% is 189 GW from 62,832 wind turbines with a yield of 390 TWh/a. Taking into account the different conditions in the various wind regions and the resulting different plant designs, an actual realizable potential of 198 GW with 410 TWh/a is found.

In the study by the Federal Environment Agency [UBA 2013], a GIS-based terrain model is used to investigate the influence of distances to settlements on the basis of spatial planning on the wind energy potential. Similar to the previous study, low wind and high wind turbines are used with a similar design.

The distance rules to settlements, roads and highways, overhead lines and all other types of land use were taken into account in a very detailed and specific manner.

The theoretical area potential consists of 13.8% of the area of Germany (Table 2.3). This corresponds to a capacity of 1190 GW with a yield of 2900 TWh/a at 2440 average full load hours. The distance to settlements of 600 m is three times the peak height of the wind turbines.

Table 2.3 shows the reduction of the potential, if the distance to residential areas is enlarged by spatial planning rules. Minimum distances of 2000 m would only allow a yield of 81 TWh/a. This means that the requirements for a predominantly regenerative energy supply in Germany could not be met by 2050. A distance of 1200 m to residential land, however, is sufficient for a yield corresponding Table 2.2 and at the same time to limit environmental impacts on settlements by noise immission and rotor shadow flicker. As a scenario for 2050 therefore in Table 2.2 an onshore wind potential of 400 TWh/a is assumed. This corresponds to 55% of the potential with a settlement distance of 1200 m. In the offshore area, 200 TWh/a are estimated. Thus, the wind energy with 600 TWh/a has a portion of 63% of the electricity generation in 2050 in Germany.

Wind energy potential in Europe (EU-28)

The spatial planning rules with regard to the expansion of wind energy are very different in the individual member states of the EU. A distinction must also be made between spatial planning on land and in the offshore area. The potential of Europe's wind energy is therefore analyzed without detailed consideration of spatial regulations.

Table 2.3 Settlement distance and wind energy potential in Germany [UBA 2013]

Distance to settlement	600 m	800 m	1000 m	1200 m	2000 m
Area potential	13.8%	9.1%	5.6%	3.4%	0.4%
Share of reference potential	100%	66.3%	40.9%	24.8%	2.8%

In the "National Renewable Energy Action Plan" [NREAP 2020], the member states (EU-27) have specified the national development targets for wind energy until 2020. The result is a capacity of 213.4 GW with yields of 494.57 TWh/a at average full load hours of 2159 h/a.

The technical wind energy potential by 2030 is 45,000 TWh/a onshore and 30,000 TWh/a offshore [EEA 6/2009]. WindEurope [WE 2017] has examined three scenarios with small, medium and high wind installations for the year 2030. In the onshore area are realizable installation of 207 to 300 GW is expected and in offshore areas from 50 to 100 GW. This means that yields from 650 to 1150 TWh/a are possible.

Up until 2050, it is hardly possible to make reliable forecasts for the expansion, as the realizable potentials and yields depend heavily on the future technological development of wind turbines. The technology of onshore wind turbines is expected to develop in the direction of low wind converters to enable decentralized energy supply close to consumers.

In the offshore sector, high outputs of around 10 MW per wind converter can be expected. The development of wind energy up to 2050 also depends on Europe's future energy strategy and can therefore only be roughly estimated.

Assuming a predominantly renewable energy supply in Europe, an onshore expansion to 2000 TWh/a at 800 GW power and an offshore expanding to 700 TWh/a at 200 GW appear necessary (Table 2.2). Wind energy will therefore provide more than half of the electrical energy in Europe in the future. It will therefore be the most important regenerative generation technology in Europe, ahead of hydropower.

2.3.3 Photovoltaics

Technological development and installation potential of photovoltaics
Several technologies are available for the modules of photovoltaic systems (Fig. 2.6). In 2016. Multicrystalline silicon had a production share of 69%,

Fig. 2.6 Production shares of PV technologies [FHISE 2018]

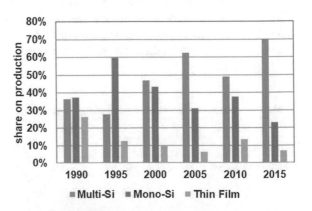

monocrystalline silicon was 25% and thin-film technology was 6%. Among the thin-film technologies, CdTe has a share of 63%, followed by copper indium selenide cells (CIS) with 28% and amorphous silicon with 9%. The energy conversion efficiency of new modules for monocrystalline silicon are 18 to 24%, in multicrystalline silicon 16 to 18%, and in thin-film technologies 12 to 16%.

In the early days of the PV, the subsidized feed-in tariffs were independent of the installation site. In Central Europe, there is now a tendency to no longer subsidize PV plants on green field [ÖSET-VO 2018], or in Germany to lower the feed-in tariffs for green space installation compared to building integration.

The roof and façade areas available for building integration of PV depend on the settlement density. In metropolitan areas, the roof area per apartment or inhabitant is smaller than that of single-family houses due to the high-rise building construction. Table 2.4 shows this for various settlement types.

Figure 2.7 shows this in a double logarithmic representation. From the point of view of the distribution grids, PV power installations in the range of 2 to 5 kW per inhabitant can mostly be integrated without low-voltage grid expansion. This applies to urban networks, which are generally well developed and have short network lengths. In rural areas, for example at resettlement farms outside a village, low grid capacities can limit the expansion of PV power. Here, high self-use rates in the building and the installation of local accumulators for storage are required.

Table 2.4 Roof area, PV power and annual energy according to settlement structures

	Population density	Roof area	PV power	Energy
	PE/km^2	m^2/PE	kW/PE	kWh/PE/a
Core area metropolitan	5000	2–10	0.4–2	380–1900
Suburban settlement	1000	10–30	2–6	1900–5700
Scattered settlement, village	50–100	15–100	3–20	2850–19,000

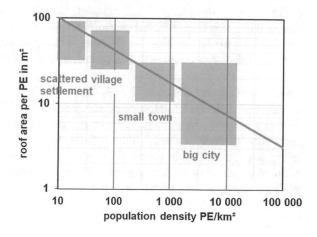

Fig. 2.7 Usable roof areas for PV per inhabitant (PE)

The physical potential of photovoltaics is of course also dependent on the local irradiation and thus on the latitude. In the northern European countries, the annual yields (full load hours) are 800 to 950 kWh/kW$_{PV}$, in central Europe 950 to 1100 kWh/kW$_{PV}$ and in southern Europe 1100 to 1700 kWh/kW$_{PV}$.

Potential of photovoltaics in Austria

Since 2015, photovoltaic systems have only been subsidized if they are installed on a building and not in a green area. This means that in the future only the potential of the roof and facade areas will be of importance. Austria has 2,191,000 buildings [Östat 2016]. Of these 87.5% are single- and two-family houses. In a research project [S4MG 2011], the potential of buildings was assessed. The areas of roofs suitable for PV are 139 km^2 and the suitable facade areas are 52 km^2. In the future, the efficiency of PV systems will be at least 20%. With average values of 950 full load hours on roofs and 650 full load hours on facades, power densities of 200 W/m^2 result.

Thus, the potential of the roof areas is 27.8 GW with yields of 26.4 TWh/a and of the facades 10.4 GW with 6.7 TWh/a. The total capacity is calculated to be 38.2 GW with 33.2 TWh/a. The value for PV generation in 2050 of 30 TWh/a given in Table 2.2 assumes, that 90% of the suitable roof and facade areas are used for the installation of PV systems. Despite the large roof area potentials, PV systems with efficiency of at minimum 20% are required.

Potential of photovoltaics in Germany

Photovoltaics is a necessary and important regenerative generation technology in Germany with around 1000 full load hours. The feed-in tariffs differentiate between systems on residential buildings, on noise barriers and other systems. Other systems are, for example, systems on open spaces that receive less funding than systems on residential buildings.

It is foreseeable that with further expansion of photovoltaics, the systems on land will be limited to open spaces for reasons of landscape protection. In the following, it is therefore assumed that installations on buildings and noise barriers will predominate by the year 2050. If an installation capacity of 3 kW per inhabitant is assumed in accordance with the potential on buildings, this results in an installation capacity of 240 GW by 2050 for Germany with approximately the constant population. In Table 2.2, a power of 250 GW with a yield of 250 TWh/a was assumed.

Potential of photovoltaics in the EU-28

To determine the potential, Europe can be divided into three climate zones. In southern Europe, the average number of full-load PV hours is 1500 h/a with a population of 30% of the EU-28 members, in central Europe 1000 h/a with 60% of the population and in northern Europe 900 h/a with 10% of the population. If 3 kW of PV installation power per inhabitant is assumed, the potential is 1500 GW. With average full load hours weighted by the number of inhabitants of 1140 h/a, the yield is 1710 TWh/a. In Table 2.2 a value of 1500 TWh/a was assumed. Here, too, the value is close to the potential limit. To achieve this target, high investments in the expansion of renewable energy are required in all member states of the EU.

2.3.4   Hydropower

Types of hydropower plants
The following types for hydropower plants can be distinguished [WKA 1997]:

- Run-of-river power plant
  - River power plant. The facilities are located in the riverbed with a turbine building and weirs. All of the river water can flow through the plant.
  - Diversion power plant. In this type of construction, a river meander is separated by a barrier structure and shortened by a branch channel. The river meander receives only a small amount of residual water and the main part flows into the power plant via the diversion channel.
- Storage power plant with natural inflow. A reservoir receives its natural inflow according to the size of the valley basin and, if necessary, by pipelines from rivers and streams from neighboring valleys. Storage power plants only have an upper reservoir with a dam. The annual natural inflow is processed via turbines and flows off via the underlying river. The energy gained in the river power plants is counted in the run-of-river power plants.
- Pumped storage power plant with natural inflow. These power plants have a lower and an upper reservoir between which the water to oscillate in the turbine and pumping operation. This operating mode is counted as a pumped storage operation, since energy from the grid is used here. Only the processing of the natural inflow is rated as hydraulic energy generation. Pumped storage power plants are also considered as storage facilities in Chap.5.

According to the European Directive 2000/60/EU *"Framework in the field of water policy"* [WRRL 2000] surface water bodies are to be returned to a good ecological status. In the case of new construction or conversion of hydropower plants, the overriding public interest in a sustainable and secure energy supply must be weighed against the resulting deterioration in the condition of the water body. The passability of the water bodies for fish migration makes the construction of fish ladders necessary. In the case of diversion power plants, residual water doping of the bypassed flow sections is planned in the medium term. This reduces the energy yield, especially in many old small hydropower plants in particular. The Water Framework Directive determines the future ecologically expansion potential of hydropower plants in Europe.

In its early days from 1880, electricity generation in Europe relied heavily on the expansion of hydropower plants. As a result, the remaining hydropower potential that is still worth expanding is limited. The future sustainable energy strategy of the EU is based therefore especially on wind energy and photovoltaics.

However, seen from the generating characteristic the hydropower represents the technically best form of renewable energy generation. It is characterized by a more uniform supply, the change of which is better predictable. Throughout the year for large run-of-river power plants a secure minimum production of about 40% of installed capacity is given. Furthermore, there are high full load hours of 4000 to 5500 h/a. The high full load hours need less grid expansion compared to land-based wind energy or to greenfield photovoltaics.

Potential of hydropower in Austria
In accordance with the study by PÖYRY [POY 2008] on hydropower potential in Austria, the total techno-economic hydropower potential is 56.1 TWh/a. Of this potential, already 5.0 TWh/a are realized by small hydropower plants (<10 MW) and 33.2 TWh/a by large hydropower plants. The remaining potential available is 17.9 TWh/a. From this 1.4 TWh/a can be achieved by optimization of existing hydropower plants. The remaining potential is 16.5 TWh/a.

Taking into account sensitive areas and the expected requirements of the Water Framework Directive, of the total remaining potential of 17.9 TWh/a, only 12.8 TWh/a appear realistically exploitable. The long-term technically and economically usable total potential of hydropower in Austria is therefore 51 TWh/a. In Table 2.2 a value of 42 TWh/a in the year 2050 is used. In Austria, in 2050 hydropower remains the dominant regenerative generation technology with a share of 37.5%.

Potential of hydropower in Germany
The technical potential of hydropower in Germany is 24.7 TW/a [WKA 1997]. In 2005, the realized generation was 19.7 TWh/a. According to the *National Renewable Action Plan* [NREAP 2020], this is to be expanded for Germany to a value of 20 TWh/a by 2020. In Table 2.2 a potential of 22 TWh/a is assumed for Germany in 2050, which can be realized by optimizing of existing run-of-river power plants and new construction of pumped storage power plants with natural inflow.

Hydropower potential in EU-28
The potential of hydropower in the European Union and its development up to 2020 are compiled in the *National Renewable Energy Action Plan* [NREAP 2020] (Table 2.5).

Large hydropower with power over 10 MW has a share of 86% and small hydropower 14%. The yield of small hydropower is expected to grow at 1.6% per year in the period from 2010 to 2020 and the large hydropower with 0.5%/a.

Table 2.5 Development of hydropower in the EU-28 by 2020 [NREAP 2020]	Year		2005	2010	2015	2020
	Power	GW	115	118	126	136
	YIELD	TWh/a	347	346	357	370
	Full load hours	h/a	3013	2930	2830	2730

In order to achieve the EU's sustainability goals by 2050, countries that still have good hydropower potential in particular will expand their hydropower use. Assuming average growth in hydro power in the EU-28 of 0.9%/a, this results in a yield from hydropower of 500 TWh/a in 2050, as shown in Table 2.2.

2.3.5   Biomass

Biomass is also referred to as a renewable raw material. It can be obtained from forestry or from agriculture on green areas. Residues materials from these areas also count as biomass. Mixed uses for the production of food (e.g., grain) and energetic use (e.g., straw) are also possible.

Agricultural biomass (energy crops, field crops).

- sugar beet
- starchy crops: potatoes, cereals, corn grain
- lignocellulose plants: miscanthus

Agricultural residues

- straw from cereals
- stalks of corn
- grasses
- shells and pulp (e.g., when harvesting pumpkin seeds)

Forest residues

- logs of wood
- forest chips
- pellets
- tree bark
- leaves

Biomass has low solar efficiency. Here, the energy content of the biomass grown by photosynthesis is considered in relation to the solar radiation energy over the growth period, based on the area for agriculture or forestry.

The biological production efficiency is between 3.5% for sugar beets and 0.1% for slowly growing biomass. Due to the large usable area for biomass and with environmentally friendly natural production, biomass will also have a high significance in the future.

Biomass can either be used directly, e.g., as firewood, or converted into solid, liquid or gaseous products. These are pellets, liquid biofuels or biomethane. The conversion processes are shown in Table 2.6.

Fermentation is a biological conversion process.

Table 2.6 Biomass, utilization processes and end products

Plant/fruit	Organic raw material	Procedure	End product
Sugar beet, potato, cereal grain, corn grain	Sugar, starch	Fermentation	Bioethanol (liquid)
Rapeseed sunflower seeds	Vegetable oil	Transesterification	Biodiesel (liquid)
Sugar beet, cereals, corn, grasses	Whole or partial plant, biogenic residues	Anaerobic fermentation	Biomethane (gaseous)
Trees, shrubs, grasses, miscanthus	Lignocellulose	Combustion	Heat, electricity
All types of biomass	All kinds	Pyrolysis and synthesis	Hydrogen Methane Methanol Heat Electricity

Esterification for the production of biodiesel is a chemical conversion process. In this process, the glycerin contained in vegetable oils or animal fats is replaced by a methanol alcohol. This leads to a lower viscosity, which comes close to diesel made from mineral oils.

Combustion is a chemical oxidation process.

Direct combustion of logs, wood chips or pellets can generate heat, electricity or both. The efficiency of biomass boilers is relatively high and is 65 to 85% [EUBB 2015]. The electricity generation from biomass as a fuel with steam generator, steam turbine and generator have low efficiencies of 15 to 25%, since at high combustion temperatures, the ash of biomass tends to melting, thus high temperatures have to be avoided. Combined heat and electricity generation in decentralized district heating plants is useful. Here, combined efficiencies of 65% for heating and 20% for electricity generation are possible, which results in a total efficiency of 85%.

From biogenic raw materials and residues, electricity, heat and gaseous energy carriers or liquid fuels can be generated simultaneously in a fluidized bed process using steam gasification. The biomass is first broken down into CO and H_2 by pyrolysis. In a high-pressure fluidized bed process with the addition of steam, a synthesis gas can be produced from this as a hydrocarbon with longer carbon chains using the Fischer–Tropsch process [Güssing 2002].

Potential of biomass in Austria

The agricultural and forestry areas in Austria amount to 75,400 km^2 and represent 90% of the land area. Of this, 13,770 km^2 is arable land, 17,900 km^2 is permanent grassland, 33,100 km^2 is forestry land and 9910 km^2 is unproductive land.

Biomass production up to the year 2050 consists of agricultural energy crops as well as agricultural residues and auxiliary materials, furthermore, from forestry wood and saw by-products. The wood harvest is subdivided into roundwood for sawing, industrial wood and firewood. In Table 2.7 and Fig. 2.8 Austria's biomass potential is shown with its mean value and the minimum and maximum values.

The average biomass potential of 91.5 TWh/a, reaches the magnitude of the renewable electricity generation potential of 112 TWh/a in 2050 of. The electricity generation from biomass in 2050 with 20 TWh/a, shown in Table 2.2, assumes that, with efficiencies for generating electricity of 20 to 35%, about half of the biomass potential is used for cogeneration.

The large biomass potential enables applications for heat supply, electricity generation, for the production of biodiesel for the mobility sector and biogas for thermal backup power plants. This can reduce the electricity demand in the sector coupling power-to-heat and power-to-mobility and thus also the necessary efficiency targets in the end use of electricity.

Biomass potential in Germany

In Germany, a *National Biomass Action Plan for Germany* [BioAP 2010] was decided. The share of biomass on the final energy consumption should be increased from 6.2% in 2007 to 10.9% in 2020.

The following measures are planned to reduce competition for use in food production on its cultivation areas:

Table 2.7   Energetic biomass potential in Austria in 2050 [KraHa 2009]

2050	Average	Minimum value	Maximum value
	TWh/a	TWh/a	TWh/a
Agriculture	44.5	42.1	46.5
Agricultural residues	9.2	7.8	17.2
Forestry	27.8	23.2	32.8
Saw by-product	10.0	10.0	15.0
Total	91.5	83.1	111.5

Fig. 2.8   Energy potential of biomass in Austria [KraHa 2009]

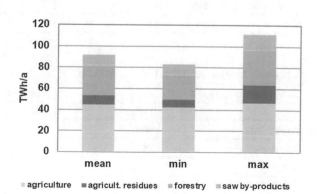

agriculture   agricult. residues   forestry   saw by-products

- Increase in yields per hectare of agricultural lands
- Tapping of residues and by-products, that do not compete with food production or other uses
- Development of wood potential
- Facilitating the establishment of short-rotation plantations

In the potential analysis, the calorific value of the primary product is used here and not that of the converted end product, e.g., ethanol or biodiesel. When considering the final energy use of biomass, the energy potentials are therefore smaller. A compilation of the results of several studies on the energy potential of biomass in Germany [BioDE 2007] yields a total potential in the range of 235 to 315 TWh/a (Table 2.8).

Other potential analyzes come to different results of 360 TWh/a [BioDE 2002] and reach up to 430 TWh [LFZ 2012].

In 2016, electricity generation from solid, liquid and gaseous biomass in Germany was 50.8 TWh/a [EEZ 2016]. An increase to 60 TWh/a in Table 2.2 appears realistic.

Biomass potential in the EU-28
The biomass potential in the European Union is shown in Table 2.9 and Fig. 2.9. Residues include agricultural and forestry residues, sewage gases and biogenic components of the waste.

A comparison of the maximum annual biomass potential per capita results in 13,500 kWh per inhabitant for Austria, 3800 kWh per inhabitant in Germany and

Table 2.8 Biomass potential in German in 2050 [BioDE 2007]

Production sector 2050	Average value	Minimum value	Maximum value
	TWh/a	TWh/a	TWh/a
Forestry	108	90	125
Activatable forest wood	38	35	40
Agriculture	52	40	65
Activatable energy crops	78	70	85
Total	275	235	315

Table 2.9 Biomass Potentials in the EU [JRC 2015]

Production sector 2050	Average value	Minimum value	Maximum value
	TWh/a	TWh/a	TWh/a
Agriculture	1800	1300	2700
Residues	300	150	450
Forestry	1300	800	2800
Total	3400	2250	5950

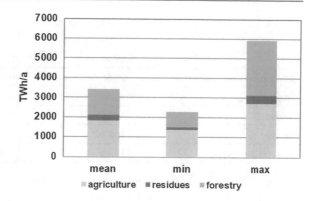

Fig. 2.9 Biomass potential in the EU in 2050 [JRC 2015]

11,600 kWh per inhabitant in the EU. The lower potential in Germany is also a consequence of the high population density (Table 2.1).

The study by Bentsen and Felby [BenFel 2012] showed a much greater range of variation and higher maximum values for the biomass potential in the EU, which is between 1 290 and 19,200 TWh/a.

Electricity generation from biomass at 300 TWh/a in 2050 (Table 2.2) corresponds to a use of 9% of the biomass potential.

2.4   Significance of Biomass in Renewable Energy Supply

The energetic use of biomass represents a significant part of the regenerative energy supply of the future. With its raw potential, it reaches the capacities of hydropower, wind energy and photovoltaics. Compared to hydropower, wind energy and photovoltaics, however, there are additional expenditure for transport, conditioning, energetic utilization and for the treatment of residual and waste materials.

Conversion losses in the production of gaseous or liquid fuels or in electricity generation reduce the usable potential of biomass, and more complex processes lead to cost increases. However, automation of processes and technological efficiency improvements in biomass utilization will make it economically viable in the future as petroleum product prices rise.

When used as a combustible material for the sector heat, as a biofuel for the transport sector or for the generation of balancing energy in thermal backup power plants, this can minimize the demand for electrical energy, needed for the sector couplings *Power-to-Heat* and *Power-to-mobility*. This makes it possible to close the gap shown in Table 2.2 between the electricity generated from renewable sources and the final electricity demand.

In summer, for example, there is an oversupply of photovoltaics and there is an increased demand for electricity in winter. This gap can be closed either by long-term storage of potential energy in the water volume of central pumped storages or

Table 2.10 Energy storage as fuel and as pumped storage content

| | Calorific value | Density | Energy content |
	kWh/kg	kg/liter	kWh/m^3
Rapeseed oil	10.97	0.92	10,092
Sunflower oil	10.94	0.92	10,068
Biodiesel	10.20	0.88	8980
Heating oil	10.80	0.95	11,260
Biomethane, liquid	13.89	0.42	5834
Pump storage, head = 500 m	–	1.00	1.4

in the form of biofuels in tanks. Central, high-capacity pumped storage plants can use large storage volumes as annual storage facilities and thus create an energy balancing. As Table 2.10 shows, the energy density of 1.4 kWh/m^3 at a head of 500 m is relatively low. But large storage capacities are economical, since the volume of storage represents only about 15% of the total investments of a pumped storage plant, compared to the costs of the dam, the pressure tunnel and the cavern with the electric generating facilities.

Solid and liquid biomass with their high energy density can manage with smaller storage volumes (Table 2.10). This makes small decentralized power plants possible, which provide local balancing in energy cells with small generation capacities and fuel storage.

As Table 2.10 shows the calorific values from vegetable oils and biodiesel are comparable with heating oil. A high energy density is achieved with around 10,000 kWh/m^3. Regionally produced bio fuels can therefore be used in the future especially in many small decentralized cogeneration units with electricity generation at high overall efficiencies.

In the future, pumped storage will be particularly necessary for short-term energy generation and the provision of control and balancing energy.

2.5   Summary

The potentials of renewable energy supply in Europe are limited. These are roughly twice the electricity demand of 2016 or 40% of the final energy demand in 2016 or 80% of the final energy demand in 2050. Around 20% of the final energy demand in 2050 must be generated by thermal balancing power plants. Therefore, a demand reduction of 50% based on the final energy demand in 2016 is necessary. About 70% of the energy savings come from new efficiency technologies such as heat pumps and electric vehicles, and 30% from changed sufficient user behavior.

By sector coupling *power-to-heat* and *power-to-mobility* a high degree of efficiency improvement and a transition to sustainable energy supply is possible.

Biomass is of great importance in sustainable energy supply. It can be used directly to generate heat and can be used in the mobility sector as a biofuel, especially in long-distance transport for quick refueling.

As long-term storable energy carrier with high energy density it can serve in biomass power plants for decentralized provision of balancing energy in and reduce the need for less economic long-term storage capacities.

Due to the limited potential of renewable electricity from hydropower, wind energy and photovoltaics, energy from biomass will be important in the future to close the gap between the temporal course of the renewable generation and the course of the energy demand.

References

[BentFel 2012]   Bentsen, N.S., Felby, C.: Biomass for energy in the European Union—a review of bionenergy resource assessments. BMC (open access publication 2012)

[BioAP 2010]   Nationaler Biomasse Aktionsplan für Deutschland

[BioDE 2002]   Hartmann, H., Kaltschmitt, M.: Biomasse als erneuerbarer Energieträger: Eine technische, ökologische und ökonomische Analyse im Kontext der übrigen erneuerbaren Energien. Landwirtschaftsverlag, Münster (2002)

[BioDE 2007]   Aretz, A., Hirschl, B.: Biomassepotenzial in Deutschland - Übersicht maßgeblicher Studienergebnisse und Gegenüberstellung der Methoden. Dendrom-Diskussionspapier Nr. 1 (März 2007)

[BWE 2011]   Potenzial der Windenergienutzung an Land - Kurzfassung. Bundesverband Windenergie, BWE (2011)

[EEA 6/2009]   Europe's onshore and offshore wind energy potential. European Environmental Association, Report EEA 06/2009

[EEZ 2016]   Erneuerbare Energie in Zahlen. Nationale und internationale Entwicklung im Jahr 2016. Bundesministerium für Wirtschaft und Energie (2016)

[EUBB 2015]   Report on conversion efficiency of biomass. Basis-Biomass Availability and Sustainability Information System (July 2015)

[EW 2014]   Das realisierbare Windpotenzial Österreichs für 2020 und 2030. Forschungsprojekt des Kima- und Energiefonds 2014, Auftragnehmer Energiewerkstatt

[FHISE 2018]   Photovoltaics Report. Fraunhofer Institut für Solare Energiesysteme, Fraunhofer ISE (2018)

[Güssing 2002]   Hofbauer, H., Rauch, R.: Biomasse-Kraftwerk in Güssing

[JRC 2015]   The JRC-EU-Times model. Bioenergy potentials for EU and neighboring countries. Joint Research Centre JRC-Report EN 27575 EN (2015)

[KraHa 2009]   Kranzl, L., Haas, R.: Strategie zur optimalen Erschließung der Biomassepotenziale in Österreich bis zum Jahr 2050 mit dem Ziel einer maximalen Reduktion an Treibhausgasemissionen. Forschungsprojekt Bundesministerium für Verkehr, Infrastruktur und Technologie bmvit, Bericht 44/2009

[LFZ 2012]   Langfristszenarien und Strategien für den Ausbau der erneuerbaren Energien in Deutschland unter Berücksichtigung der Entwicklung in Europa und global. Schlussbericht DLR, Fraunhofer IWES und IfnE (2012)

[NREAP 2020]   Beurskens, L.W.M., Hekkenberg, M.: Renewable Energy Projections as Published in the National Renewable Energy Action Plans of the European Member States. European Environment Agency (2011)

[ÖSET-VO 2018]   Ökostrom-Verordnung 2018, vom 22. Dezember 2017, Bundesgesetzblatt. 408. Verordnung der Republik Österreich

[Östat 2016]   Statistik Austria: Gesamtenergiebilanz Österreich 1970–2016 Überblick

[POY 2008]   Wasserkraftpotenzialstudie Österreich, Pöyry-VEÖ (2008)
[S4MG 2011]   Super-4-Microgrid—Nachhaltige Energieversorgung im Klimawandel.
Forschungsprojekt des Klima- und Energiefonds (2011)
[TUW 2014]   Enengel, S., Steiner, H., Marsch, V.: Der Einfluss der Raumplanung auf die
Windkraft. Masterprojekt TU Wien (2013/2014)
[UBA 2013]   Potenzial der Windenergie an Land. Studie zur Ermittlung der bundesweiten
Flächen- und Leistungspotenziale der Windenergienutzung an Land. Umwelt-Bundesamt,
Berlin (2013)
[WE 2017]   Wind Energy in Europe: Scenarios for 2030. Wind Europe (2017)
[WKA 1997}   Giesecke, J., Mosonyi, E.: Wasserkraftanlagen—Planung; Bau und Betrieb.
Springer (1997)
[WRRL 2000]   Directive 2000/60/EC of the European Parliament and of the Council of 23
October 2000 establishing a framework for Community action in the field of water policy

Energy Economy of the Efficiency

3

3.1 Development of Generating Costs and Grid Tariffs Until 2050

The cost of electricity in residential, industrial or in commercial end-use during the transition to a predominantly renewable electricity supply is of great importance to household affordability or industrial and commercial competitiveness.

The electricity tariff for customers will in future consist of the following components:

- generation costs of the individual regenerative energy sources
- costs for storage and feeding back into the grid
- generation costs for balancing energy in thermal backup power plants
- grid tariffs for more expanded transmission and distribution networks
- metering and billing costs
- taxes and duties

In the case of regenerative energy sources, a reduction in investment costs can be expected with large production volumes. Table 3.1 shows this for photovoltaic systems for the period from 2010 to 2050. Here, an interest rate of 3% and annual maintenance costs of 4.5% of the investment value are applied uniformly.

In the case of onshore wind turbines, it is assumed that low wind turbines will be used in future onshore far away from the coast. They have larger rotors with generator outputs in the range of 3 MW. This means that reinforced towers and foundations are necessary. The price reduction is therefore lower and a decrease from 1000 €/kW in 2020 to 900 €/kW in 2050 is assumed. In the offshore area, a decrease from 4000 €/kW to 3000 €/kW is assumed.

Table 3.2 shows the development of the generation prices for the different renewable energy sources adjusted for no inflation on the basis of 2016.

© Springer Fachmedien Wiesbaden GmbH, part of Springer Nature 2022
G. Brauner, *System Efficiency by Renewable Electricity*,
https://doi.org/10.1007/978-3-658-35138-0_3

Table 3.1 Development of the generation costs of photovoltaic systems

Year	2010	2020	2030	2040	2050
Investment €/kW	1500	800	600	500	400
Full load hours	1000	1000	1000	1000	1000
Service life a	20	23	25	28	30
Annual costs €/kW	168	85	62	49	38
Electricity price €ct/kWh	16.8	8.5	6.2	4.9	3.9

Table 3.2 Development of the generation costs of renewable energy sources

€ct/kWh	2010	2020	2030	2040	2050
Photovoltaic	16.8	8.5	6.2	4.9	3.9
Wind onhore	5.1	4.8	4.4	4.1	3.9
Wind offshore	11.7	10.7	9.7	8.8	7.7
Hydropower	3.0	3.0	3.0	3.0	3.0
Biomass	5.0	5.0	5.0	5.0	5.0
Geothermal energy	5.0	5.0	5.0	5.0	5.0

Fig. 3.1 Generation costs of photovoltaics and wind energy

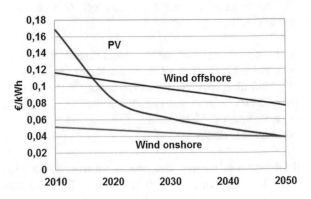

Figure 3.1 shows the generation costs of photovoltaics and wind energy. The costs for the integration of renewable energy sources are not included in the generation costs, since, according to the organization of the electricity market, these shares are to be included in the consumer network tariffs.

The development of electricity prices up to the year 2050 can be calculated at full costs from the shares of the individual renewable generation types in Table 3.2. The analysis is limited to the Austrian and German electricity markets.

When considering the German electricity market, Chap. 2, Tab. 2.2 assumes that the share of renewable electricity from baseline in 2016 will linearly increase until 2050 to the specified end value there. The respective energy price can be calculated using a share matrix for the energy mix and the cost development according to Table 3.2. In the generation prices, the thermal backup power plants and the pump storage must also be taken into account with their development up to 2050.

To balance the electricity system, thermal power plants with decreasing and pump storage with increasing shares are required. In 2020, this will represent 65% of total electrical energy in Germany. A share of 3% from pumped storage and a share of 62% from thermal power plants are assumed. In Austria, due to the high share of hydropower, only 30% of the annual energy is generated by backup and pumped storage power plants in 2016.

In Austria, only combined cycle gas and steam power plants are existing since 2020 as thermal power plant types. The operating hours of this type of power plant will probably be reduced from 2500 full load hours in 2020 [VDE 2012] to only 1000 h/a in 2050. At the same time, it is assumed that the gas price will increase from 2.5 €ct/kWh to 5.0 €ct/kWh in 2050.

In Germany, the last nuclear power plant will be shut down in 2022. The thermal power plants consist of coal and combined cycle gas fired power plants. In the long term, there is also a trend towards gas fired combined cycle power plants only. Figure 3.2 shows the development of the generation costs of pumped storage and combined cycle power plants.

The service life of combined cycle power plants is about 20 years and the investment costs are about 1300 €/kW. Due to the priority of renewable energy, the combined cycle power plants operate with decreasing full load hours. This causes an increase in the share of fixed costs in the feed-in tariffs.

As was shown in a study for Austria [S4MG 2011], a complete renewable energy supply is currently not feasible, since this requires annual storage capacities with large volumes and a short usage time. Neither with regard to environmental effects of large storage capacities nor from an economic point of view is this

Fig. 3.2 Generation costs of combined cycle power plants and pumped storage

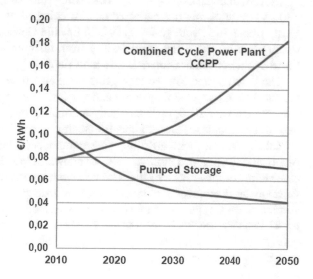

reasonable. Therefore, in 2050, around 15 to 20% of the annual energy must be provided from the combined use of pumped storage and thermal power plants.

For pumped storage, the specific investment costs of 1000 €/kW are assumed. This is a mixed value that that takes into account both, new facilities including reservoirs, as well as the expansion of existing pumped storage capacities by only exchanging turbines and generators for higher power and making new pressure tunnels. The energy price for pumping is set to 3 €ct/kWh (Fig. 3.2 upper curve) and 0 €ct/kWh (Fig. 3.2) lower curve). Pump storages are expected to increase from 1000 operating hours in turbine operation in 2016 (about 2000 h in pumping and turbine mode) to 2500 h per year in turbine operation in 2050 (5000 h in total) due to the expansion of renewable energies. This will reduce the fixed costs and thus the specific feed-in tariffs. This effect is very pronounced, when pump storages use renewable surplus energy at its marginal cost of zero cents for pumping and releases it as peak balancing energy in the event of a shortage. They can then be operated very economically. The increasing full load hours of pumped storage and the falling hours of the thermal backup power plants represents in future a competitive advantage for pumped storage. Moreover, this prevents that surplus renewable energy must be switched off and the later on resulting generation gaps can be filled by pumped storages instead from thermal power plants.

Stored regenerative energy can thus be used more cost-effectively than energy from thermal power plants. This makes pump storage economical in the future.

The hours of use (full load hours) of the electric energy grid will decrease from currently 5500 h/a to only 2000 h/a in Austria in 2050 due to the increased use of renewable energy [S4MG 2011]. Higher power at lower full load hours has to be transmitted. The grids must therefore be strengthened for higher transmission capacities at lower usage hours. This also applies to Germany and the EU-28.

In the transmission grid by converting old overhead lines from 220 to 380 kV, the transmission capacity can be enlarged at least by a factor of three. Furthermore, high voltage DC (HVDC) links in Germany from north to south are required, to transport the wind energy from offshore and northern onshore wind farms to the consumer centers in southern Europe.

The medium and low voltage grids must also be extended for their capacities for the integration of PV systems and for fast-charging stations for electric vehicles. By the year 2050, approximately a doubling of distribution grid capacities is necessary. Also, taxes and charges for electricity will roughly double.

Taking into account all these cost components, including the backup power plants and the pumped storage used, as shown in Fig. 3.3, the full-price tariffs for electricity for customers in the low voltage network will develop. Excluding the German tariff element for development of renewable energy (renewable energy surcharge, EEG), the price of electricity is expected to increase from 22 €ct/kWh to 37 €ct/kWh by 2050, according to the share matrix of different generating sources and of the increased grid tariffs including future grid extension and for new taxes and duties to be expected.

Fig. 3.3 Development of electricity prices in Germany and Austria

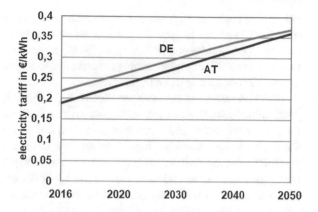

Fig. 3.4 Development of electricity price and PV generation costs

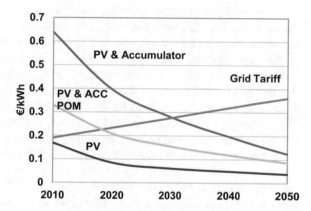

Austria starts at a lower price level because of the high proportion of hydro-power. Due to the significant expansion of photovoltaics and wind energy, grids and pumped storages, will bring electricity tariffs in line with the German level by 2050.

Price increases by inflation are not taken into account. It can be assumed, that when prices rise, wages also rise, so that only the scaling shifts but affordability remains unchanged.

3.2 Photovoltaics and the Public Ownership Model (POM)

Figure 3.4 shows the development of the generation costs of photovoltaic systems. In the case of centrally installed green-field systems, the entire energy is fed into the grid and can only be marketed via the networks and thus with grid tariffs.

In the case of decentral installed PV systems, the energy can be used directly in the building and only a small part is fed into the grid. Figure 3.4 shows the generation costs for own use in house installations without a grid tariff. The lowest curve (PV) applies to systems without an accumulator, the top one with an accumulator. The middle one (PV & ACC POM) represents a PV system with an accumulator, where the surplus energy is marketed to neighbors in the house according to a public ownership model (POM).

If PV systems on the roof or facade are owned by private individuals, their profitability results from the avoidance of purchase costs from the grid. The difference between the grid electricity price and the cost of self-generation represents the profit. As described in Chap. 7, using time series analysis of the solar radiation, small PV systems without accumulator, which in its power are adapted to the electricity demand of the end user, can on average only be used to about 35% in their feed-in, since the load characteristic of the demand of households and of the PV generation over the year do not permit a higher coverage.

As a result, the specific costs for owners are higher than shown in the lower curve in Fig. 3.4. The profitability can be improved by selling the surplus energy. However, since many PV systems will be connected to the grid at the same time in the future and the solar radiation will reach all these systems at the same time, only low remuneration can be expected when exporting energy via the gird. So, the yield will be near to the marginal costs, which are near to zero for PV. Therefore, the production costs for a use of 35% are therefore more than twice as high as the generation costs shown in Fig. 3.4. Nevertheless, the ownership use of PV systems is more economical than purchasing energy from the grid.

If there is a stationary accumulator with the PV system, that is designed according to economic criteria (Chap. 7), the self-use can be increased to 50%. The price of the accumulator is currently approximately 700 €/kWh at a service life of 6 years and will change to around 300 €/kWh at a service life of 10 years in 2050. Until the year 2030 PV systems with accumulators will be competitive (Fig. 3.4).

The legal frame of energy regulation of the electricity market in Germany and Austria is now changed, so that public ownership models are possible and that excess energy of PV inside a house can be sold to neighbors. This does not require an external energy provider and accordingly no grid charge has to be paid, because of the electricity transport is exclusively within the house. This is called a public ownership model.

The owners of the PV system can then pass their surplus energy at their full cost to neighbors. So public ownership models are already economically today if the dimensioning is made according to the demand of a house, as shown in Fig. 3.4. The gap between rising grid tariffs and falling system costs thus suggests increasing profitability of PV systems based on the public ownership model.

3.3   Efficiency and Affordability

Efficiency is not only necessary to adapt the energy demand to the available renewable potential, but also to enable affordability. As has already been shown, the renewable energy supply results in higher electricity prices.

Table 3.3 Annual costs of an inefficient conventional household

	2016	2050
Electricity demand kWh/a	4000	4000
Electricity price	0.22 €/kWh	0.37 €/kWh
Annual electricity costs €/a	*880*	*1480*
Living space m²	100	100
Heating energy requirement kWh/m²/a	150	150
Oil heating price €/kWh	0.08	0.15
Heating costs €/a	*1200*	*2250*
Car mileage km/a	7000	7000
Liters/100 km	6.5	2.5
Fuel price €/liter	1.3	3.0
Travel costs €/a	592	525
Total costs €/a	*2672*	*4255*

Table 3.4 Annual costs of an efficient, innovative household

	2016	2050
Electricity demand kWh/a	2000	2000
	0.22 €/kWh	0.37 €/kWh
Annual electricity costs €/a	*440*	*740*
Living space m²	100	100
Heating energy demand kWh/m²/a	20	20
Electricity tariff €/kWh	0.22	0.37
Heat pump coefficient of performance	4	4
Heating costs €/a	*110*	*185*
e-Car mileage km/a	7000	7000
kWh/100 km	15	15
Travel costs €/a	230	388
Total costs €/a	*780*	*1313*

Traditional inefficient consumption behavior with unchanged or increasing demand leads to high energy costs. This is illustrated by the example of a household with two persons, shown in Table 3.3. The rising costs for electricity, heating oil and fuels will noticeably increase future energy costs.

For the same living space and the same mobility demand, Table 3.4 shows the energy demand and the costs for a household that was already efficient in the past. Through measures such as use of efficient home appliances and new lighting electricity, demand can be reduced by half for general demand. The heating energy demand can be reduced through thermal insulation of the building and the use of a heat pump. When using an electric car, a significant reduction in the energy demand for mobility is still possible despite rising electricity costs.

Even taking into account the higher energy costs in 2050, the efficient household here has total energy costs of about half of that compared to an inefficient household in 2016 or 30% compared to 2050. Figure 3.5 shows the development of energy costs in both households. Rising energy costs will make efficiency measures particularly attractive in the future. The energy turnaround is affordable with efficiency measures and, at the same time, they enable to comply with the potential limits of renewable energy.

Efficiency will therefore be the most important measure for the success of the energy transition in the future. The renewable targets cannot be achieved without efficiency. The higher energy costs will give impetus to greater efficiency in the future.

The depreciation of investments in efficiency measures was not taken into account here. It is assumed that investments are not made solely on the basis of efficiency improvements, but that efficient devices are only procured after the end of service life of the existing inefficient ones. It is also assumed, that in future the depreciation on investments in new, efficient devices will roughly correspond to the depreciation costs of the devices to be replaced.

3.4   Energy Economic Assessment of Efficiency

Efficiency measures are assessed according to costs and benefits. Measures are economical if the costs of the energy saved are higher than the annual depreciation on investments for energy efficiency. This represents a business analysis.

The annuity method is predominantly used for the assessment of energy efficiency, which considers annual depreciation on the capital employed as well as annual maintenance and operating costs. Operating costs include expenses for energy, auxiliary materials, insurance, taxes and labor costs.

Fig. 3.5   Energy cost comparison of a conventional inefficient and an innovative efficient household

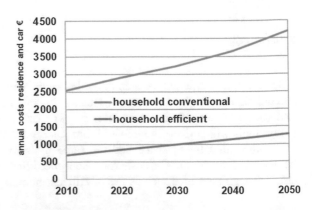

From the perspective of companies, efficiency can be economically interesting not only through cost savings but also through an improvement in image, for example, an award as eco-hotel can increase the number of overnight stays and so increase revenues. However, this non-technical "detour profitability" shall not be considered here.

Annuity Method

With the annuity method, an investment A_0 is made at the beginning and should be repaid in n periods in equal installments Z until the remaining debt B_0 becomes zero.

$$B_0 = A_0 - \sum_n Z \equiv 0 \tag{3.1}$$

$$Z = \alpha_n \cdot A_0 \tag{3.2}$$

The annuity factor represents the portion of the capital raised that is repaid annually.

$$\alpha_n = \frac{(q-1) \cdot q^n}{q^n - 1} \tag{3.3}$$

The interest factor q results from the interest rate. (e.g., 3%, $q = 1.03$).

If the number of repayment periods n corresponds to the service life of a measure, then the asset is just amortized at the end of service life.

An efficiency measure is cost-effective, if the saved annually amount of energy ΔE at the specific costs k_E has a higher value than the annual installment Z.

$$Z = \alpha_n \cdot A_0 \leq k_E \cdot \Delta E \tag{3.4}$$

If the annually saved costs of energy are viewed as an increased repayment rate, a shortened repayment time in m periods can be calculated from this, which is considered the **amortization time** or **payback time**.

$$Z_m = \alpha_m \cdot A_0 = k_E \cdot \Delta E \tag{3.5}$$

The payback period m can be determined iteratively with this equation.

Table 3.5 shows the payback period for efficiency measures in the household. On the one hand, incandescent lamps of 60 W (IL 60 W) are being replaced by compact fluorescent lamps of 11 W (CFL 11 W) or LED lamps, and other energy efficiency measures are being investigated. A uniform interest rate of 5% is assumed.

The costs for energy-saving lamps in the staircase are higher here, since they are switched on up to 100 times per day, while apartment lamps are only switched on once a day. Unlimited switch-on lighting is therefore used in the staircase.

If the amortization time of a measure is in the order of magnitude of the service life, there is no economic benefit. Since energy prices will rise in the future, it may still be appropriate. As energy costs rise in the future, payback times will decrease.

Table 3.5 Energy payback time for efficiency measures in the household

Area	Measure	Replacement	Substituted power (difference)	Usage time	Invest costs	Costs of substituted energy	Amortization
			kW	h/a	€	€/kWh	years
Staircase light	CFL 11 W	IL 60 W	0.049	2000	15.5	0.22	0.95
Apartment light	CFL 11 W	IL 60 W	0.049	2000	7.0	0.22	0.36
Apartment light	LED 8 W	IL 60 E	0.052	2000	9.0	0.22	0.39
Shower	Solar thermal	Oil heating	33.6	1000	33 600	0.062	30.0
Window	Isolating window	Single glass pane	0.14	3500	500	0.065	11.0
Heater	Thermo—static valve	Simple valve	0.11	3500	250	0.07	3.0
Swimming pool	2 New pumps	2 Old pumps	4.0	4000	3200	16,000	0.8

Fig. 3.6 Contracting for PV
systems

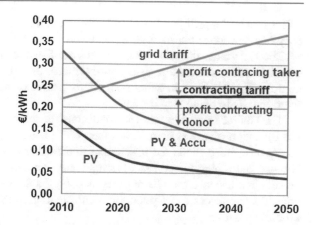

Table 3.6 Price and lifespan of PV systems and accumulators up to 2050

	2010	2020	2030	2040	2050
PV system prices in €/kW	1500	800	600	500	400
PV system service life in years	20	23	25	28	30
Accumulator prices in €/kWh	700	600	500	400	300
Accumulator life in years	6	7	8	9	20

3.5   Contracting

Generation Contracting

Photovoltaic systems are suitable for generation contracting. As Fig. 3.6 shows, the prices for PV systems and accumulators are falling. In the future, the electricity tariffs from the grid will increase as a result of the grid expansion, the provision of reserve power plants and pump storages and the expansion of renewable energies. This opens up a gap between purchase prices from the grid and costs of self-generation from photovoltaics.

Electricity from PV systems without accumulators already has lower self-generation costs than the grid tariff. This will also apply in future to systems with accumulators. Table 3.6 shows the likely development of prices and service life until 2050.

In the case of generation contracting for PV, a long-term contract of 15 years, for example, is concluded between the contract donor (CD) and the contract taker (CT). The CD hereby undertakes to finance, build and operate the PV system. At the end of the contract period, the system is depreciated and becomes the property of the CT, who can continue to use the system until the end of its service life. Between the CD and the CT a price for the use of PV-generated electricity is agreed that is lower than the grid purchase tariff for electricity (Fig. 3.6). The difference to the grid purchase tariff represents the monetary benefit of CT, which may be relevant for tax purpose in the future.

The CT grants the CD the use of the roof or facade area for PV by entering an easement in the land register. Furthermore, the CD can market the excess PV energy. This revenue can either be partially passed on to the CT or used as revenue from the CD to depreciate the asset.

The contract also contains the conditions for maintenance and repair as well as the operation of the system. From the point of view of the CT, he saves himself the need to raise funds for investment and further expenses for operation, maintenance and repair. No contract arrangement as a member of a balance group will be required because the CD appears as energy aggregator of several PV systems, and thus the access of its pools of small plants into the electricity market is cost efficient possible.

Efficiency Contracting
Efficiency contracting makes it possible to reduce the demand in end use of energy through efficiency measures. This form of contracting is often found in the area of heating systems. For example, replacing an old oil heating system by a more efficient condensing heating system or a heat pump result is a lower heating energy demand. Some of the savings can be passed on to the CT to reduce heating costs. The rest is used to cover depreciation of capital and operating costs for the new facility and provides a return on investments for the CD.

3.6   Efficiency Measures and Monitoring

The energy efficiency can be improved by the following measures:

Energy efficiency directive as a top-down approach [EU 2012/27 EC]
Efficiency measures are determined for the public sector, industry and energy suppliers. The energy savings are to be evaluated in audits and reported annually to a monitoring agency. Annual percentage targets are set for efficiency improvements. Penalties are provided for violations. Households and small business do not have to carry out monitoring themselves and are exempt from penalizations.

Efficiency measures as a bottom-up approach
This includes several measures that are intended to lower the energy demand in the end use.

- **Ecodesign Directive**. This defines minimum standards for efficient products and devices, e.g., the average energy demand for a standardized usage process or the stand-by demand. Devices that do not meet these minimum standards may not be produced in the EU, nor may be imported. This measure is only effective in the longer term, as the lifetime of most devices is around 20 years.

- **Labeling** according to energy consumption classes in accordance with Directive 30/2010 EU. Devices are compared in their efficiency with reference specifications. The best devices receive the label A, the worst G. Since the efficiency of the devices has been increased through technical progress, labels A ++ to D were required. The scale must be regularly redefined and reset to the highest label A, here since 2021 the classification returned to the range of A to G.
- **Disposal** of old devices or device replacement campaigns. By setting incentives, old devices are exchanged for efficient new equipment, with a discount for new equipment as an incentive act.
- **The energy efficiency certificate** of buildings shows the specific energy demand of a building for heating and air conditioning and gives tenants the opportunity to assess the costs required for this.

Incentive financing, subsidies and advertising will bring about the use of efficient devices and awareness among end users of energy sectors for efficiency measures.

Sectors for efficiency measures according to the EU Energy Efficiency Directive [EU 2012/27 EC] are:

Public Sector
The public sector should act as a model for energy efficiency measures. The annual renovation rate of public buildings with more than 250 m^2 should be at 3% per year. In general, local energy efficiency plans must be drawn up and energy management systems introduced. Products, services and buildings with high energy efficiency standards are preferred and efficient consumption behavior is to be created.

Energy efficiency guidance should provide advice on efficient consumption behavior and record the annual improvement through audits and monitoring.

Energy Supply
For utilities, national energy efficiency goals are defined. The energy suppliers, who are actually responsible for the delivery of energy services, have to achieve an annual reduction in the energy demand of their end customers in accordance with national requirements. This represents a conflict of interest. If the end customer collective does not fulfill the requirements, penalties are provided for the energy suppliers.

Furthermore, individual measuring devices (smart meters) are to be introduced for the time and the actual energy consumption. The construction and retrofitting of power plants with power ratings exceeding 20 MW is subjected to obligations for cost analysis. This should lead to the construction of efficient power plants.

Combined heat and power (CHP) generation is to be introduced on a mandatory basis. Grid tariffs are to be designed in such a way that energy savings and consumption control become possible. National energy efficiency standards shall be established for generation plants. Table 3.7 shows the improvement in

Table 3.7 Efficiency and emissions from thermal power plants [VEO 2004]

Power plant type	CO_2 thermal	CO_2 electric	Efficiency electric	Efficiency with CHP
	kg/kWh$_{th}$	kg/kWh$_{th}$	%	%
Gas and steam	0.20	0.33	62	75–85
Hard coal new	0.35	0.76	45	70–80
Brown coal new	0.56	1.10	45	70–80
Hard coal 1960	0.35	1.10	33	–
Brown coal 1960	0.56	1.80	27	–

efficiency and the reduction in CO_2 emissions from thermal power plants since 1960 [VEO 2004].

The thermal power plants have achieved high levels of efficiency thanks to the technological development of the power plants and the promotion of CHP.

Industry and SME

Industry and SMEs (**S**mall and **M**edium **E**nterprises) have a significant share of the total energy consumption from 25 to 30%. This sector has made good progress in the past with an average of 1.5%/a in energy savings. Particularly in energy-intensive industry, efficiency measures are economical to implement. Companies with 50 or more employees must set up an energy management system or carry out energy efficiency audits. There is an obligation to report via the energy accounting and savings monitoring. "Best practice" should contribute to the dissemination of energy-efficient technologies. Industry benchmark parameters are helpful here. In the context of benchmarking, energy demand parameters are determined per production unit or per service. This allows companies with similar products or services to be compared with one another. For hotels these are, for example, energy parameters in kWh per overnight stay or in bakeries kWh per product weight produced. Using these characteristic values, the production technology can then be optimized, e.g., in bakeries, the baking surfaces can be loaded with different products at the same time instead of working sequentially over longer heating periods. Energy efficiency can therefore not only be improved through technological measures, but also through organizational ones.

Household

In the household sector, no concrete measures or controls on energy efficiency are planned. Indirect measures are used to try to promote efficiency. The EU ecological design directive [EU 2009/125 EC] represents a bottom-up approach. Herewith minimum-efficiency standards are prescribed for the individual product groups, which are used in the home, e.g., for washing machines, heating and air conditioning units, lighting and televisions. Products that do not meet these minimum requirements, may either be produced nor imported in the EU.

The energy suppliers should use a top-down approach to lower the households' energy demand through indirect measures. This includes the introduction of the smart meter and customer billing, with which the change in energy demand compared to the last billing period can be clearly shown. Efficiency advice and device replacement campaigns represent further measures to improve efficiency.

The reduction in electricity demand in the household must be reinterpreted in the future. On the one hand, the introduction of efficient devices results in a reduction in demand. On the other hand, the demand for electricity increases if, for example, an oil heater is replaced by an electric heat pump or a gasoline operated car with an internal combustion engine is replaced by an electric vehicle. These substitution potentials should be taken into account in the future—possibly on a lump-sum basis. Because in the long term, the electricity demand will increase due to the substitution of fossil energy with renewable electricity.

Ecological Design Directive 2009/125/EC
The directive creates an integrated product policy that encompasses the entire ecological life cycle of a product, from raw materials to production, use and disposal. Material and energy flows are assessed in terms of their environmental impact over the entire life cycle. The aim of the guideline is to achieve efficient products with a low environmental impact. The goal is to reduce energy consumption, in particular electricity consumption over the life cycle. The Ecological Design Directive is to be implemented into national legislation with minor national adjustments. Special guidelines apply to the individual product groups.

3.7   The Rebound Effect

Rebound refers to the effect that energy savings, for example, lead by new efficient technologies to cost savings, which in turn bring the procurement of more or more efficient products, which in the end result in identical or higher energy demand.

Figure 3.7 shows rebound effects for televisions. The short bars correspond to televisions with a power requirement according to the EU energy label of efficiency class A and the long bars correspond to class G. If an old TV set with a cathode ray picture tube of 26″ (CRT TV) and efficiency class G is replaced by a new TV set with LCD flat screen with a diagonal of 65″ of the most efficient class A, the power consumption remains the same. A less efficient LED screen has even an additional demand.

The rebound effect can be found in many areas of energy end use. Examples for this are:

- Moving from a small apartment with a poor thermal insulation standard to a thermally insulated detached house with a large living space.
- Replace a car that uses a lot of fuel with a more efficient one and drive the new vehicle more frequently.

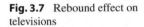

Fig. 3.7 Rebound effect on televisions

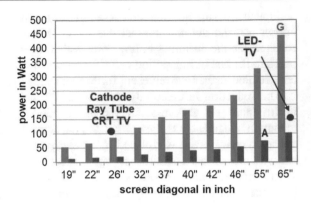

- Replacing a small car with low motor efficiency by an SUV with modern motor for the mobility in a big city.
- Replacing a light bulb with a chandelier with lots of LED bulbs.
- Replacing an old inefficient refrigerator with a new class A one and reusing the old device as a wine cooler in the cellar.

3.8   Gray Energy and Energy-Payback-Time

Gray energy denotes the amount of energy that is required for production, transport to the end customer, installation there, and disposal. The life cycle assessment of a combi refrigerator consists of around 20% energy for production and transport (gray energy), 80% energy for cooling operation, 0.6% for maintenance and repair and 0.1% for disposal [SAFE 2005].

From an economic point of view, the replacement of an old, inefficient device only makes sense, if the grey energy of the new device is less than the energy savings through higher efficiency over its service life, compared to the old device to be disposed of.

The "energy-payback-time" describes the period of time in which the gray energy of the new device is saved due to the higher efficiency of the new device compared to the old one. Exchanging of an old device is only appropriate in terms of energy if:

$$E_{grey,new} \leq \Delta E_{old-new} \cdot T_{new} \tag{3.6}$$

The energy difference per unit of time due to the operation of the new device in comparison to the old device, added up over the service life T_{new} must be less than the gray energy of the new device $E_{grey,new}$.

Figure 3.8 shows the replacement of refrigerators of different years of construction by a new efficient refrigerator. For its production an energy of 900 kWh was required and the operating energy demand has only a value of 227 kWh/a.

The operating energy demand of the old refrigerators decreases with the year of construction from 1000 kWh/a when in built in 1970 to 300 kWh/a when built in 2000. For the old refrigerators with year of construction 1990 and older, replacement is energetically expedient. From the year 2000 onwards, refrigerators with Label A have a mean annual energy demand of less than 300 kWh/a. Because of the small difference between the operating energy demand an exchange of refrigerators since 2000 leads to long payback times and is not appropriate from an economic perspective. In the case of devices with a low label D to G, on the other hand, the replacement can be energetically appropriate.

3.9 Analysis and Monitoring of Energy Efficiency

An analysis of the energy demand is a prerequisite for being able to carry out efficiency measures economically. Improving efficiency is a continuous process, as new technologies and production processes will continue to improve efficiency potential in the future.

In the energy-intensive industry such as metal production and processing, the cost of procurement of energy, can account for up to 40% of the gross sales. In the paper industry as well as the glass, ceramics, stone and earth industries, between 20 to 30%, and in the chemical industry about 20%. In these companies, efficiency measures in production are an essential part of the corporate strategy. The adaptation of the processes requires a high level of specific knowledge and is therefore carried out by in-house specialist departments and specialized manufacturers of the production equipment. External advice on energy efficiency is only required in rare cases and can only be provided by specialized consulting.

In contrast, there is a great need for energy advice in the sector of small and medium-sized enterprises (SMEs) and in the households. SMEs are divided into production, services and trade.

Fig. 3.8 Payback time when replacing refrigerators

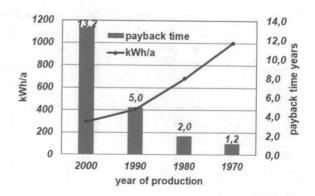

Small business for the production of goods (selection)

- bakery
- brewery
- mill
- food production
- agriculture
- clothing
- metal processing
- electrotechnical goods and equipment

Service business (selection)

- office
- hairdresser
- restaurant
- hotel
- installation
- repair shop

Trade (selection)

- retail trade
- Wholesale
- Supermarket

Table 3.8 shows typical characteristic values of the energy demand of the different types of small and medium business found in different brochures for SMEs [WKO 2012]. This makes it possible to find the types of application of energy use with the greatest efficiency potential.

Process heat and process cooling make up the largest share in the bakery and brewery.

Table 3.8 Typical characteristic values of the energy demand of SMEs

	Room heating (%)	Room cooling (%)	Process heat (%)	Process cooling (%)	Lighting (%)	Machines & office (%)	Warm-water (%)
Bakery	8	–	45	25	6	6	10
Brewery	5	10	20	40	5	15	5
Office	30	10	–	–	29	30	1
Hair stylist	55	20	3	–	3	5	14
Restaurant	15	3	30	17	15	10	10
Hotel	31	15	5	3	12	17	17
Supermarket	4	5	5	40	35	10	1

Example supermarket

In the supermarket, process cooling and lighting are considered here in terms of efficiency potential. Either centrally supplied refrigerated shelves or decentralized plug-in freezers can be used to cool the food [Peritsch 2006]. In the case of the central cooling shelves, refrigerant lines are laid to the shelves and heat exchangers are located outside the building. The advantage is that no heat is released into the sales area. The disadvantage is that the coolant lines worsen the efficiency.

The decentralized freezers have a higher degree of efficiency, but additional measures with cooling surfaces suspended from the top are required for room cooling, which worsens the overall degree of efficiency. Efficient, newly built supermarkets therefore have central cooling systems and use daylight openings in the ceiling to reduce the energy demand for lighting.

Example hotel

In the following, the basic procedure for an energy audit is presented for a hotel. Table 3.9 [BerVoss 2012] (with additions) can be used to assess energy efficiency in the hotel sector. The different categories from luxury to three-star hotels have different characteristic values, which for an average hotel occupancy of 40% apply.

The annual costs of energy compared to sales are typically 5 to 7%. Since the margin at the hotel industry is only about 4% to 6%, efficiency improvements can significantly increase the return on sales.

To analyze the efficiency of a hotel, top-down approaches are best suited, in which the annual income, the annual energy costs and the hotel occupancy are considered. These data are easy to obtain. Synthetic bottom-up models, which are based on the installed devices and the facilities of the final energy demand, are much more complex and error-prone, since they do not include user behavior.

Table 3.9 Characteristic values of hotels for efficiency benchmarking [BerVoss 2012]

Reference	Energy	Dimension	Luxury	*****	****	***	Target value
Overnight stay (o)	Electricity	kWh/o	38	20	10	5	<3
	Heat	kWh/o	40	21	16	7	<4
Room	Area	m^2	60	40	20	10	<20
	Electricity	kWh/room/a	12,000	6000	2000	600	<300
	Heat	kWh/room/a	13,000	7000	3500	1000	<500
Hotel area	Electricity	kWh/m^2/a	195	160	90	60	<30
	Heat	kWh/m^2/a	200	170	160	100	<50
Overnight stay (o)	Water	liter/o		522	308	250	<240
Energy/sales			6–12%	5%	5.9%	6.8%	

With the data collected in this way, a comparison with characteristic values of the corresponding hotel category according to Table 3.9 is possible. The average energy demand of the hotel industry is a comparison with the state-of-the-art, but does not yet show any potential for efficiency.

Efficiency potentials can be identified and logged through a walk-through inspection:

Electricity

- Lighting: are incandescent lamps or halogen lamps still in use?
- Do the rooms have key card switches for presence detection?
- Do the rooms have minibar refrigerators or are drink vending machines installed at the corridor (4-star and luxury hotels require a minibar)?

Water/hot water

- Shower heads with limited consumption (6–12 l/min)
- Single lever mixer with economy setting
- Two-button system for toilet flushing
- Solar thermal warm water generation (saves up to 85% of energy, but cost–benefit analysis required)

Heating

- Radiators cleaned and vented regularly
- Thermostatic valves on the radiator
- Individual temperature control per room
- Night reduction of the flow temperature
- Speed-controlled circulation pump in the heating system
- Condensing boiler, pellet heating or heat pump
- Window contact switches off radiators (winter hotels)
- Ventilation system with heat recovery

Cooling

- Avoid heat first instead of cooling
- Increase setpoint in summer (6° temperature difference to the outside)
- Presence detection through key card switch
- Thermal insulation of the building envelope
- Insulating window
- Thermal insulation curtain
- Bivalent heat pump (geothermal heat–geothermal cooling)

A ranking of the possible measures can be carried out on the basis of these parameters. The most effective measures, which are those with the highest energy savings at the lowest cost, should be implemented first. In an improvement process lasting several years, the individual measures can then be implemented one after the other and an annual efficiency improvement can be reported to the national monitoring agency for energy efficiency. A detailed guide for energy management in the hotel and catering industry can be found in [WKO-ÖHV 2009].

3.10 Energy Management

The European Energy Directive [EU 2012/27/EC] calls for the introduction of energy management systems. Step-by-step instructions [EA-KLI 2007] have been developed by an international team. The energy management represents a continuous improvement process, which should lead to a reduction of the energy demand and thus the energy costs. It consists of the following sub-processes:

- Business case

In the preparatory phase, knowledge is acquired for the investigated sector of energy application or the specific process. A project team that is suitable for analysis, implementation planning and evaluation is put together, according to the technical requirements.

- Energy audit

Benchmarking is carried out as part of an energy audit. Benchmark parameters are useful to enable a classification in comparison to "best practice" processes. Based on an analysis of the *status quo*, efficiency potentials can be identified and possible efficiency measures put together.

- Energy Action Plan

The possible efficiency measures are evaluated and ranked in terms of their feasibility and the expected costs and economic benefits. From this, an energy action plan is created, which is worked out in detail.

- Energy coordination

The action plan is coordinated with the energy coordinator and modified if necessary. The internal staff responsible for energy and, if necessary, external contractors coordinate the tasks and initiate the orders.

- Implementation

The energy efficiency measures are implemented in accordance with the energy coordination.

- Evaluation

The effectiveness of the measures is evaluated and readjusted if necessary.

- Revision and restart of the process

The energy efficiency measures are revised and measures already taken are modified or expanded or new measures are started according to the ranking.

3.11 Summary

The generation costs for energy will rise as a result of the large-scale construction of renewable generation plants, the expansion of storage systems, the necessary expansion of energy grids and the need for backup generation plants for a secure energy supply. The increase in energy and grid costs can be synthesized on the basis of the cost development of the regenerative generation plants, the degree of expansion of the grids and the costs of the backup supply. A full cost model is used as this allows the energy system to operate economically. The use of marginal cost models is not appropriate, since the marginal costs of regenerative generation systems are zero. The similar renewable supply characteristic of wind and PV synchronize the generation, which is why periods of surplus generation and low prices cannot be offset by periods of generation shortage and high prices.

In the field of photovoltaics, generation contracting will be attractive in the future due to falling system costs and rising energy prices. In a longer-term contractual commitment, the generating plants can be installed and operated by the contracting provider. The contracting party receives locally generated energy at low cost without taking grid tariffs into account.

Efficiency measures must be economical. *Rebound effects* should be avoided and the *energy pay-back time* can be used as an indicator of the expediency when replacing devices.

Energy audits and energy management system are necessary in the future for a continuous efficiency improvement process. In the long term, they can increase profitability in companies and among private users and promote suitable efficiency measures.

References

[BerVoss 2012]   Bernhard. S., Voss, K.: Energieverbrauch in der Hotellerie. EU Wuppertal (2012)

[EA-KLI 2007]   AEA-Handbuch: Schritt für Schritt Anleitung für die Implementierung von Energiemanagementsystemen. Austrian Energy Agency und klima:aktiv (2007)

[EU 2009/125 EC]   Directive 2009/125/EC of the European Parliament and of the Council of 21 October 2009 establishing a framework for the ecodesign requirements for energy-related products

[EU 2012/27/EC]   Directive 2012/27/EU oft the European Parliament and of the Council of 25 October 2012 on energy efficiency, amending Directive 2009/125/EC and 2010/30/EU and repealing Directives 2004/8/EC and 2006/32/EC

[Peritsch 2006]   Peritsch, M.: Supermärkte als Energiezentralen. Berichte aus Energie- und Umweltforschung 2/2006. Forschungsprojekt des Klima- und Energiefonds

[SAFE 2005]   Ökobilanz Kombi-Kühlschrank Elektrolux ERB3105. Schweizer Agentur für Energieeffizienz (2005)

[S4MG 2011]   Super-4-Micro-Grid—Nachhaltige Energieversorgung im Klimawandel. Forschungsprojekt des Klima- und Energiefonds (2011)

[VDE 2012]   Erneuerbare Energie braucht flexible Kraftwerke. Studie des VDE (2012)

[VEO 2004]   Abschätzung der Verfügbarkeit der Erzeugungskapazitäten in Österreich bis 2015 und deren Auswirkungen auf die Netzkapazitäten. Studie TU Wien im Auftrag des Verbandes der Elektrizitätsunternehmen Österreichs (2005)

[WKO-ÖHH 2009]   Energiemanagement in der Hotellerie und Gastronomie. Leitfaden. bmwfi, WKO und ÖHV (2009)

[WKO 2012]   Energiekennzahlen, Energiepotenziale und Tipps (for different SME's: bakery butcher, hairdresser and others). Wirtschaftskammer Österreich (2012)

Efficiency and Sufficiency

4

4.1 Importance of Efficiency

The energy efficiency describes the ratio of the given benefit to the necessary energy input to achieve it. Efficiency is primarily related to generation plants or to plants or devices for the end use of energy. From the point of view of the overall system, it is appropriate to also include the infrastructures of the transmission and distribution networks and the energy storage in the consideration of efficiency. Furthermore, including the coupling of the heating and gas supply sectors with the electricity system can enable a more holistic view of energy efficiency. This is known as system efficiency. Energy efficiency is therefore a very complex area of work in research, development and implementation. The definitions of energy efficiency are correspondingly varied.

4.2 Generation Efficiency

Generation efficiency refers to the conversion efficiency of active energy sources such as power plants or regenerative power converters in the form of wind, PV or hydro systems.

Thermal generation plants
Generation efficiency of thermal power plants $\eta_{th.PP}$ based on the fuel flow:

$$\eta_{th.PP} = \frac{P_{el}(1 - e_b)}{h_u \cdot \dot{m}} \tag{4.1}$$

With

P_{el} electrical power output of the generator in kW
e_b own demand in per unit (5% corresponds to 0.05 pu)

© Springer Fachmedien Wiesbaden GmbH, part of Springer Nature 2022
G. Brauner, *System Efficiency by Renewable Electricity*,
https://doi.org/10.1007/978-3-658-35138-0_4

h_u lower calorific value of the fuel in kWh/kg
$\dot{m}$ fuel flow in kg/h

The efficiency of steam power plants has reached values of 45% through higher steam pressures and temperatures as well as through reheating. With combined heat and power, overall efficiencies of 75 to 92% are possible.

Combined cycle power plants (CCPP) as a combination of a gas turbine, whose waste heat is used to generate steam for a steam turbine, achieve efficiencies of up to 62% [Siemens 2014]. The thermal power plants have to meet new requirements with a predominantly renewable energy supply [VDE 2012].

- Due to the priority of renewable energy generation, thermal power plants have the task of closing generation gaps quickly and reliably. In the so-called dark doldrums (dark lulls), when there is insufficient regenerative generation from photovoltaics and wind energy, they should also enable a high level of security of supply in the longer term.
- In the past power plants were optimized for high efficiency at their nominal power rating. New requirements are high levels of efficiency even in the partial load range.
- Because of the fluctuations in regenerative generation, thermal power plants must be able to generate high generation gradients. They will have a higher number of starts and shut-downs and they should also allow for low minimum load. Modern steam turbines can be operated in the power range from 16 to 100%.
- After the shutdown of a power plant, a rapid restart (hot start) should be possible. In the case of new design of combined cycle power plants, the restart time can be reduced from around an hour to 40 min [Siemens 2014].

Table 4.1 shows the development of the efficiency of thermal power plants [Schaeff 1977].

Gas engines with outputs of up to 20 MW enable the construction of decentralized combined heat and power plants. They allow start times of 5 min. Large power gradients are possible with several parallel gas engines. A first power plant of this type went into operation with the Kiel coastal power plant in 2018 (Table 4.2) [Kiel 2018].

In purely electric summer operation, the efficiency with 45% is lower than in a combined cycle power plant with 62%. In winter operation with heat extraction, the overall efficiency is comparable to 90%. The short start times, even after long periods of standstill (cold start), and the ability to vary the starting power using the number of engines starting at the same time, make this power plant very flexible.

Fuel cells
In fuel cells, the energy content of the fuel is converted into electrical energy with the help of an oxidizing agent. As a gaseous fuel, hydrogen is oxidized with oxygen. Butane or methane can also be used in special fuel cells. Methanol is possible as a liquid fuel.

Table 4.1 Development of the efficiency of thermal power plants [Schäff 1977]

	Construction year	Steam pressure bar	Steam tempera-ture °C	Reheater	Efficiency
James Watt steam engine [Mar 2002]	1776	1.5	115	No	3%
Parson	1888	8	175	No	10%
Double expansion	1900	12	300	No	15%
Coal-fired power station	1950	210	525	Single	31%
	2000	250	525	Double	43%
	2010	300	600	Double	45%
Combined cycle power plant [Siemens 2014]	2014		1200/450		62%
Gas engine	2018				52%

Table 4.2 Gas engine power plant Kiel [Kiel 2018]

Gas engines	20×10 MW	Quick start capability	≤ 280 s
Overall power		Gross efficiency	
Heat	192 MW	Thermal	45%
Electricity	190 MW	Electric	45%

Fuel cells generate direct current and require a converter to operate on the grid. They consist of electrodes that are separated by an electrolytic ion conductor. The electrolyte is a semipermeable membrane, and is depending on the type of fuel cell permeable for protons or other ions. The following types of fuel cells are known (Table 4.3)

- Alkaline Fuel Cell (AFC)
- Proton Exchange Membrane Fuel Cell (PEMFC)
- Direct Methanol Fuel Cell (DMFC)
- Phosphoric Acid Fuel Cell (PAFC)
- Molten Carbonate Fuel Cell (MCFC)
- Solid Oxide Fuel Cell (SOFC)

The alkaline fuel cell (AFC) does not allow CO_2 in the oxygen for oxidation. It can therefore not be operated with air. However, in the water electrolysis by renewable surplus energy, hydrogen and oxygen can be stored separately and thus used as pure gases for the AFC (Table 4.3).

Table 4.3 Temperature and efficiency of fuel cells

Type	Electrolyte	Fuel (anode)	Oxidation (cathode)	Ion in the electrolyte	Temperature °C	Efficiency
AFC	Potassium hydroxide	H_2	O_2	OH^-	150–220	40–60%
PEMFC	Polymer membrane	H_2	O_2	H^+	10–100	35–60%
DMFC	Polymer membrane	CH_3OH	O_2	H^+	80–100	40–50%
PAFC	Phosphoric acid	H_2	O_2	H^+	110–200	35–40%
MCFC	Alkali carbonate melt	H_2, CH_4	O_2	CO_3^{--}	550–1000	50–60%
SOFC	Oxide ceramics	H_2, CH_4	O_2	O^{--}	600–1000	50–60%

The polymer electrolytic membrane fuel cell, also called proton exchange membrane fuel cell (PEMFC) is less sensitive, allows a lower and wider temperature range and can be operated with air as the oxidizing agent. It is therefore used in vehicles with fuel cells and as a stationary house system or small power plant to generate electricity and heat.

Fuel cells produce due to their catalytic oxidation no nitrogen oxides. They have no noise emissions and only produce water (PEMFC) and carbon dioxide (DMFC, MCFC, SOFC) at the exit. They are very environmentally friendly. Further advantages are easy controllability via the fuel supply and the quick start capability.

> *The disadvantage in the regenerative environment is that the efficiency of hydrogen electrolysis is only about 60 %. This reduces the overall efficiency from hydrogen generation and electricity generation to 25% to 35%.*

Wind turbines
Generation efficiency of the wind turbines η_{Wind} related to the wind available:

$$\eta_{Wind} = \frac{P_{el}(1 - e_b)}{\frac{1}{2}\rho \cdot A \cdot v^3} \tag{4.2}$$

With

P_{el} electrical power output of the generator in kW
e_b own demand of the wind turbines in per unit
ρ air density in kg/m^3
A rotor area of the wind turbines in m^2
v wind speed in m/s

Wind turbines can theoretically use up to 59% of the kinetic energy of the wind. In reality, efficiencies of up to 45% are achieved.

So far, inland wind turbines have been designed for a specific power (ratio of generator power to rotor area) around 400 W/m².

Low wind turbines have an enlarged rotor area. According to Table 4.4, the design is in the range from 200 to 270 W/m². This enables them to achieve their rated output even at lower wind speeds and achieve up to 3000 full load hours inland instead of around 2000 full load hours.

With a limited generator power $P_{el.Wind}$ of around 4 MW, they generate a greater amount of energy E_{Wind} to the grid over the year according to their higher full-load hours T_W.

$$E_{Wind} = P_{el.Wind} \cdot T_W \tag{4.3}$$

The reduced generator power in relation to the mechanical rotor power of low wind turbines causes less need for transmission power extension and is thus cost saving in network expansion. From this it becomes clear, that not only the efficiency of the renewable generation plants, but also the efficiency of the entire system of generation and grid must be considered.

Figure 4.1 shows the relationship between rotor diameter and nominal power of normal wind turbines (blue) and low wind turbines (orange). Onshore there is a tendency towards low-wind turbines with nominal power of 4.5 MW and rotor diameters of 140 to 160 m, with hub heights in the range of 120 to 165 m.

The trend of the development is towards nominal powers of 10 MW in the offshore area. A design similar to that of normal wind turbines is chosen here.

Photovoltaic systems
Generation efficiency of PV systems in relation to the solar supply:

$$\eta_{PV} = \frac{p_{el}}{g_{el}} \tag{4.4}$$

With

p_{el} electrical power per PV area in kW/m²
g_{el} global radiation around the location of the PV system in kW/m²

The efficiency of PV modules in production has reached values of 24% for mono crystalline silicon (Mono-Si), 20% for multi crystalline silicon (Multi-Si), 17% for

Table 4.4 Typical characteristic values of normal and low wind systems

Rotor diameter	Rotor area F_R	Generator power P_G	P_G/F_R	Hub height
m	m²	MW	W/m²	m
Normal wind turbine				
110	9500	4.0	420	100–140
Low wind plants				
140	15,390	4.0	260	140–160
155	18,870	4.0	212	140–160

Fig. 4.1 Rotor diameter and
nominal power of normal and
low wind systems for offshore
and onshore installation

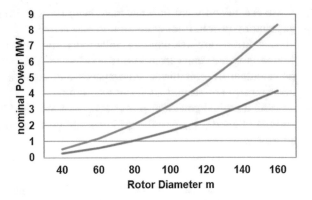

Fig. 4.2 Loss of efficiency
of PV modules

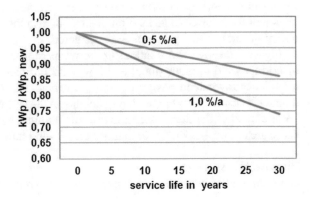

amorphous silicon (A-Si) and 15% for CdTe thin-film modules. The degradation
of the modules is between 0.5%/a and 1%/a. This enables service lives of 30 years
with efficiency losses of 15 to 25% (Fig. 4.2).

The collector efficiency as the ratio of the generated electrical power to the
power of the global radiation, based on the area of the collector, is also not the
only selection criterion. The energy input for the production of a PV plants is also
important in comparison to the generated energy over service life. Furthermore,
the effort for grid integration, storage and maintenance has to be considered.

4.3   Usage Efficiency

The efficiency of usage or functional efficiency refers to the efficiency of passive
devices that receive energy and convert it into another form of energy, for example
in building heating or process heat, light, mechanical work or radiation for communication. The ratio of intended output to energetic input describes the efficiency
of end use here.

In the case of lighting devices, the ratio of the luminous flux generated to the electric power used is considered in lumen per watt. From the point of view of an end user, however, the illuminance on an illuminated work surface in lux as lumens per square meter is also important. In this way lighting systems with reflectors are preferable to such, which radiate in a large spherical angle. It depends on the intended use, whether the entire luminous flux or the surface radiation is chosen as a criterion.

With direct electrical heating, the efficiency is close to 100%. According to their coefficient of performance, heat pumps can convert up to four times the amount of heat per electrical energy input.

4.4   Operational Efficiency

The operational efficiency is the efficient utilization of a process for providing energy services or for the production of material goods (e.g., energy consumption per manufactured good).

In case of energy services, for example, the demand for energy in stand-by mode can be reduced by temporarily switching off escalators and having them switch on again only when requested by people sensors.

In production processes a baking oven with a higher degree of filling has a higher operational efficiency and thus a lower energy demand per bakery product. Thermal insulation of a baking oven therefore only brings high technological efficiency. With appropriate logistics, high operational efficiency can also be achieved.

In thermal power plants, decoupling of electricity and heat output can be achieved by heat storage. In the event of peak electrical power demand, the unused waste heat is fed into a large heat storage tank with a capacity for several hours and shifted to the period with high heating demand.

4.5   Resource Efficiency

Resource efficiency refers to a low demand for resources in the field of energy supply.

In terms of *material*, this means a small amount of material or energy is required for the production or operation of energy converters, grid systems or in end use.

In terms of *environment* impacts on the landscape and emissions are evaluated, for example, fewer routes for transmission lines, fewer wind energy converters, fewer hydraulic transverse structures for run-of-river power plants, or fewer emissions from thermal energy conversion processes or from energy use in production processes are required.

Energy efficiency in end energy demand is always associated with resource efficiency in the field of energy conversion and transmission, since lower energy demand is connected with fewer energy converters and less grid expansion.

In the area of end use, on the other hand, there is usually a higher demand for resources, for example, for the thermal insulation of buildings, the production and use of heat pumps and the production and recycling of accumulators.

Here mainly the generation efficiency, the usage efficiency as well as the operational efficiency are to be considered.

4.6   Life Cycle Analysis (LCA)

Life cycle analysis (LCA) is a systematic analysis of a product from production through use to disposal and recycling. During production, all expenses of raw and auxiliary materials, including energy expenditures in production, are evaluated. In addition to a material flow analysis, an energy balance is also drawn up. This includes all energy expenses for the exploration and extraction of raw materials, converting them into semi-finished products and finally producing the final product and usage, disposal or recycling.

Figure 4.3 shows the modules of this analysis process. In each module the influent material flow (M) and the energy (E) used, as well as the effluent residual material (R) and the energy loss (V) are considered. For each module also environmental impacts, both for influent end effluent material and energy flows must also be considered. The horizontal arrows represent transport processes that are assigned to the next stage as energy demand, residual material and emissions.

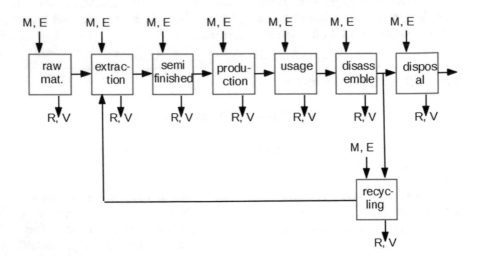

Fig. 4.3 Life cycle analysis of products

The result of an LCA is the total energy used over the usage cycle of a product, as well as the environmental impact, e.g., through the associated CO_2 emissions and the residual material balance.

Furthermore, all energy expenditures during operation are also taken into account on the basis of material flow analyzes and their energetic evaluation. ISO 14044 is a standard for the life cycle assessment of the accumulated energy and material flows.

LCA analyzes are very complex and are therefore often only used for the one-off analysis of prototypes. As a result, the cumulative energy expenditure over the service life of a product is often found. Furthermore, the CO_2 emissions balance can also be considered. With the increasing development of regenerative energy supply, the regenerative energy used has an ever smaller share in the CO_2 emissions. Therefore, only the energy used over the life cycle of a product is considered here.

4.7   Harvest Factor and Harvest Time in Relation to Grey Energy

4.7.1   Definitions

The harvest factor h is the ratio of the energy generated by a system or component over its service life to the energy expenditure for its production, operation and recycling. This is also referred to as *Energy Return on Energy Invested, ERoEI or EROI*. In the case of systems for energy conversion or electricity generation, the amount of electricity generated is valuated with the cumulative energy expenditure over the service life.

Generation harvest factor h_e, amount of energy generated E_R, to energy input E_I.

$$h_e = \frac{E_R}{E_I} \tag{4.5}$$

End-use harvest factor h_n, Energy saved E_S to expenditure of energy E_I.

For end-use equipment or components, the amount of energy saved is considered over the energy expenditure.

$$h_n = \frac{E_S}{E_I} \tag{4.6}$$

The energy saved is determined by comparison with a less efficient end use. The energy invested is then the energy expenditure plant.

4.7.2   Harvest Factor of Generation Plants

Generation harvest factor of thermal power plants
Thermal generation plants use fossil resources, biogas or biomass as fuel. The harvest factor of thermal generating plants $h_{e,therm,total}$ in relation to the total energy input is

$$h_{e,therm,total} = \frac{P_n \cdot T_m \cdot n}{E_I + E_W \cdot n + P_n \cdot T_m \cdot \frac{n}{\eta} + E_R} \qquad (4.7)$$

With

P_n installed generating capacity (nominal power)
T_m number of full operating hours (= annual energy/nominal power)
n service life in years
E_I amount of energy to produce the power plant
E_W annual amount of energy for operation and maintenance
E_R amount of energy for disposing (recycling)
η efficiency of the power plant

The annual amount of energy for operation and maintenance consists of the energetically assessed demand for auxiliary materials for desulphurization and denitrification, lubrication and furthermore all energetic expenditures that are necessary for operation. All annual maintenance and repair expenses are to be taken into account as an annual average.

In this definition, the harvest factor is less than unity, as the gray energy and the energy of the fuel used is always greater are than in the grid supplied amount of energy.

Often, the harvest factor of thermal power plants uses the energy fed into the grid in relation to the gray energy for production, operation and recycling.

$$h_{e,therm,grey} = \frac{P_n \cdot T_m \cdot n}{E_I + E_W \cdot n + E_R} \qquad (4.8)$$

This results in a harvest factor greater than one and, depending on the operation time (service life) of the power plant, it can reach values of around 40 to 50.

Equation (4.7) is the complete representation of the harvest factor. However, in this representation, the gray energy for production is masked by the amount of energy in the fuel. Therefore, the result of this equation converges to the value of the thermal generation efficiency with a long service life of a power plant. The harvest factor according to this formula is smaller than one, because only part of the energy of the fuel can be converted into end energy in a thermodynamic cycle. The production harvest factor therefore has a value between 0.1 and 0.4. Nevertheless, the efficiency of thermal power plants is very high and is 42 to 47% for steam power plants and 62% for combined cycle power plants. With combined heat and power, efficiencies of 80 to 90% can be achieved.

If Eq. (4.7) were also applied to regenerative generation plants, the power of the inflowing wind would have to be used in wind energy plants. Wind energy converters for generating electricity have an efficiency of up to about 45%. In the case of PV systems, the degree of efficiency related to the solar supply is only 12 to 24%. Even with regenerative generation plants, the harvest factor would then be significantly less than one according to this formula.

In order to better evaluate the share of gray energy for the construction of a power plant, the influence of the fuel consumption and the share for the operation, maintenance and disposal are ignored. They can thus be better compared with regenerative generation plants.

$$h_{e,therm} = \frac{P_n \cdot T_m \cdot n}{E_I} \qquad (4.9)$$

According to this formula, harvesting factors up to 50 are achieved. Power plants have a high harvest factor in relation to the gray energy required for their construction, if they can be used all year round, i.e., if they have a high number of full operating hours T_m. Furthermore, a long service life over n years is favorable.

Generation harvest factor of renewable generation plants

The renewable generation plants can be divided into thermal, hydraulic, solar or aeolian (wind) plants. Thermal plants using biogas or biomass can be treated like thermal power plants. The other renewable generation plants use solar energy, wind energy, hydropower or geothermal energy. The fuel component in Eq. (4.7) or the regenerative supply is not assessed. The generation harvest factor of the renewable plants $h_{e,reg}$ is

$$h_{e,reg} = \frac{P_n \cdot T_m \cdot n}{E_I + E_W \cdot n + E_R} \qquad (4.10)$$

With

P_n installed generating power (nominal power)
T_m number of full operating hours (= annual energy/nominal power)
n service life in years
E_I amount of energy for producing the regenerative generation plant
E_W annual amount of energy for operation and maintenance
E_R power amount for disposal (recycling)

The annual amount of energy for operation and maintenance consists of the energetically assessed need for auxiliary materials, e.g., for lubrication, and furthermore of all energetic expenditures that are necessary for operation and maintenance. In the case of renewable systems, too, the share of energy for operation and maintenance as well as for disposal is often neglected and Eq. (4.9) is used.

Examples for harvest factors of generation

Table 4.5 shows some examples of the harvest factors of generation plants. For wind energy, an E-82 turbine with a power of 2.3 MW at 97 m hub height is considered. For its production, a primary energy demand of 2880 MWh is necessary [ENER 2011]. This wind converter is studied for three different locations—inland, near to the coast and direct on the coast. Here it becomes clear that the harvest factor for the renewable plants depends not only on the energy required for

Table 4.5 Harvest factors of generation plants

Plant type	Service life	Full load hours	Harvest factor
	years	h/a	
Coal-fired power plant, $\eta = 45\%$ [EROI 2013]	25	4000	29
Combined cycle power plant, $\eta = 60\%$ [EROI 2013]	25	4000	28
Small run-of-river power plant [EROI 2013]	100	3000	50
Large run-of-river power plant	50	5000	42
Wind turbine. E82, onshore [ENER 2011]	20	2200	35
Wind turbine, E82, close to the coast [ENER 2011]	20	2550	41
Wind turbine, E82, coast [ENER 2011]	20	3200	51
PV system, mono-silicon	30	1000–1800	12–20
PV system, multi-silicon	30	1000–1800	15–25
PV system, amorphous silicon	30	1000–1800	20–32
PV system, CdTe—thin film	30	1000–1800	29–48

production but also on the wind energy potential available. For all generation types in Table 4.5, the harvest factors differ due to differences in the full load hours and the average service life.

In the case of PV systems, there is a trend towards longer service live of the collectors of 30 years instead of the previous 20 years. Furthermore, the efficiency of systems from production was increased, e.g., for monocrystalline silicon from 14% (2008) to 24% (2018). In southern Europe, thin-film technologies already achieve harvest factors of around 50 at 1800 full-load hours, which are comparable to wind turbines. Photovoltaics is therefore an efficient generation technology in the longer term and has left behind the low harvest factor of the initial period, which were lower than 10.

4.7.3 Harvest Factor of End-Use Devices

For facilities and devices for the end use of energy, the harvest factor is determined by the energy saved over the service life compared to a less efficient device. If the devices to be compared have different service life, the proportion of gray energy must be taken into account accordingly. The harvest factor $h_{e,util}$ results from:

$$h_{e,util} = \frac{\Delta P \cdot n \cdot T_m}{E_I} = \frac{(P_1 - P_2) \cdot T_{L2}}{E_{I,2}} \tag{4.11}$$

With:

P_1 net power of the inefficient system
P_2 net power of the more efficient system
T_{L2} service life time of the more efficient system
$E_{I,2}$ energy for production and installation of the more efficient system

It is assumed here that an old inefficient plant is replaced by a new more efficient one in the long term over several service life periods T_{L2}. The share of gray energy avoided of the replaced old system is then only of minor influence and does not have to be taken into account.

Example of an end-use harvest factor: replacement of incandescent lamp with energy-saving lamp.

An incandescent lamp of 60 W with a lifetime of 1000 h is replaced by a compact fluorescent lamp (CFL) of 11 W with a service life of 10,000 h. The gray energy for the production of an incandescent lamp is 0.5 kWh and for a CFL 1.5 kWh (Osram). This does not take into account the energy required to obtain the raw materials and manufacture the semi-finished products.

$$h_{e,util} = \frac{\Delta P \cdot n \cdot T_m}{E_I} = \frac{(P_1 - P_2) \cdot T_{L2}}{E_{I,2}} = \frac{(60\,\text{W} - 11\,\text{W}) \cdot 10{,}000\,\text{h}}{1.5\,\text{kWh}} = 327$$

The differential end-use harvesting factor shows that in lighting, a small share of gray energy can release great savings potential. With 1.5 kWh, the new system has a larger proportion of gray energy than the replaced incandescent lamp with 0.5 kWh per unit. But since the incandescent lamp only has a lifespan of 1000 h, the proportion of gray energy over 10,000 h of use with 5 kWh is significantly larger than the 1.5 kWh of the long-life CFL.

4.8   Energy Payback Time (Energy Amortization Time)

The energy payback time or energy recovery time refers to the time period in which an energy conversion system has generated the same amount of energy, which is required for their manufacturing, operation and disposal. Often, to gray energy does not take into account the higher energy content of the fuels, because of limited efficiency in generating plants, but only the energy output is taken:

$$T_{amortisation} = \frac{grey\,energy}{yield\,per\,year} = \frac{E_I + E_W \cdot n + E_R}{P_n \cdot T_m} \text{ [in years]} \quad (4.12)$$

In many cases, the energy expenditure during the operating time is not taken into account, but only the gray energy for the production of a system is considered.

Table 4.6 shows the energetic payback time of different generation plants. In the case of thermal power plants with fossil fuels, this is a non-genuine energy payback period, since the fossil fuels are used. Here, it is a measure of the energy-efficient manufacturing, installation and operation of thermal power plants.

In the case of regenerative generation plants, these plants can in future be manufactured using regenerative energy. A physical "payback" then takes place with the regenerative energy that is generated.

Figure 4.4 shows the *payback time* of photovoltaic systems. The values given by the IEA [IEALCA 2015] from 2008 have been converted for higher efficiencies of the PV modules. The values apply to Central Europe with 1100 kWh/kW$_p$. In Southern Europe with solar radiation of 1800 kWh/kW$_p$ the values in Fig. 4.4 would be reduced to 60%.

4.9   Sufficiency and Affordability

Sufficiency refers to limiting consumption to what is needed and thus a reducing the final energy demand or the consumption of raw materials and the associated environmental impact. Examples of this are not to own a car in the big city, because there is an efficient public transport available. This does not mean that in general to waive on a car when its use is necessary, such as for long-distance transport or in sparsely populated areas with inadequate local traffic offer. However, ownership is not absolutely necessary, *car sharing* or the use of rental vehicles also count as sufficiency.

Sufficiency is also possible in the area of living if the living space is adapted to the needs, e.g., moving into a smaller apartment as a single person and thus saving energy costs and rental expenses.

Table 4.6   Energy payback period

Plant type	Full load hours	Amortization
	kWh/kW in h/a	months
Coal-fired power plant, $\eta = 45\%$ [EROI 2013]	4000	10
Combined cycle power plant, $\eta = 60\%$ [EROI 2013]	4000	10
Run-of-river power plant [EROI 2013]	3000	24
Run-of-river power plant	5000	14
Wind turbine, Vestas V112 [VES 2011]		8.0
Wind turbine, Enercon E82, on land [ENER 2011]	2200	6.8
Wind turbine, Enercon E82, coastal [ENER 2011]	2550	4.9
Wind turbine, Enercon E82, coast [ENER 2011]	3200	4.7
PV system, mono-silicon, Central Europe	1000	29
PV system, mono-silicon, Southern Europe	1800	16
PV system, multi-silicon, Central Europe	1000	24
PV system, CdTe thin film, Central Europe	1000	12

Sufficient behavior is often driven less by insight into the limited resources and energy efficiency but more by affordability and the available resources, e.g., the quote of available living space.

Sufficiency and efficiency in the area of housing

In the arear of housing, there is often a shortage of affordable living space in large cities. The living space per inhabitant therefore decreases with the size of the settlement. The mean value in sparsely populated areas is around 45 m² per inhabitant and in metropolitan areas 38 m², as Fig. 4.5 shows [Payk 2010]. The average apartment size is similar in Germany, Austria and Switzerland. It has 70 m² and 1.6 inhabitants, which is about 45 m² per person. In single-family houses, the average living space is 55 m² per person. The mean value of living area has increased since 1960 from 20 m² per person of a large family, to 45 m² in 2018 due to increasing number of single dwellings [StatDE 2010]. Due to the tendency to move from the countryside to the city and the associated higher rental prices here, it is assumed that by 2050 the living space per inhabitant will not exceed 50 m². This value is used in Chap. 8 to analyze the potential and energy demand for space heating.

The tendency, to move to a smaller living space in the city also creates efficiency through sufficiency. In rural regions, housing is predominantly single-family homes, which have a higher heating energy demand than apartment buildings due to the large outdoor surface. In multi-family houses, only about a quarter of the apartment envelope is an exterior surface, while all other surfaces lead to neighboring dwellings with a similar temperature level and therefore increase efficiency.

Sufficiency and Efficiency in the area of Mobility

In 2015, the motorized individual transport (MIT) in Germany accounted for 86% of total traffic volume [UBM 2018]. The number of cars per 1000 inhabitants depends on the gross domestic product and thus on prosperity [Stephan 2015].

Fig. 4.4 Energy-payback-time of photovoltaic systems

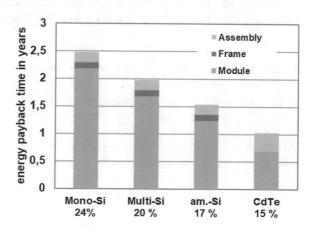

In industrialized nations, a saturation value of 550 cars per 1000 inhabitants is viewed as the upper limit in the long term [BBVA 2012].

The car density is very much influenced by the population density. In large cities with well-developed local public transport, private transport reaches its limits. Frequent traffic jams result in long MIT travel times compared to public transport. The limitation of the parking space requires long search rides and thereafter wide paths to the destination or corresponds to high parking costs. By dismantling parking areas or parking restrictions, some large cities are trying to avoid private motorized traffic in the core area and are instead relying on public transport and the development of bicycle traffic.

This makes motorized vehicles unattractive in large cities. It is not the insight of the road users, but rather the lack of opportunity that slows down MIT. Figure 4.6 shows the car density as a function of the number of inhabitants per community class. Big cities like Berlin only have 280 cars per 1000 inhabitants.

In rural regions with villages and small towns, attractive public transport with short cycle times is hardly economically feasible. Here households often have two or more cars and the vehicle density reaches the limit of up to 550 cars. In Central Europe there is an average car density of 450 cars per 1000 inhabitants.

In the future, the small electric vehicle can revolutionize suburban and rural local transport. In particular, if automated driving with moderate speed for high security in traffic is possible, also *car-sharing* can provide a flexible mobility for all—especially for elder people. As a result, the car density can be 500 cars per 1000 inhabitants in the long term. This target value with electric vehicles is used as the basis for the potential analysis of the energy supply for electric mobility up to 2050 (see Chap. 9).

4.10 Summary

Due to the limited renewable potential and the higher costs of renewable energy supply, efficiency in the generation and end use of energy are of great importance. Efficiency depends on the technology and the renewable potential. Renewable

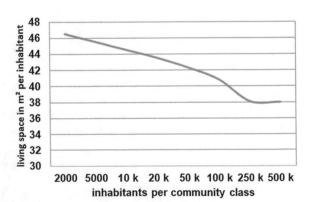

Fig. 4.5 Living space per inhabitant by community class

Fig. 4.6 Number of cars according to community class

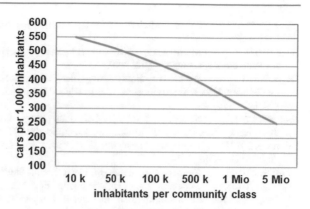

generation plants should therefore be installed at locations with a sufficiently high potential and their technology should be designed specifically for the site.

Wind generation facilities should be constructed on land preferable as low wind turbines. These systems have larger rotors with a limited generator power around 4 MW. As a result, the generation characteristics are more energy-oriented and less power-oriented with up to 3000 full load hours. Due to the smaller generator power, smaller transmission capacities are necessary and a higher level of acceptance for the grid expansion can be expected.

By now more efficient production processes are used for photovoltaic systems. Together with the increase in the efficiency of the modules, future harvest factors will be possible in southern Europe, which are comparable to wind turbines. In addition to wind energy, PV represents the regenerative technology with the greatest growth potential in the future.

Sufficiency describes the limitation of the energy demand in the end use. In all areas of end use of energy, in the area of construction, housing and mobility initiatives for this purpose are available. The urbanization leads to a higher share of the population in dense residential areas. Due to the high rental charges, this leads to smaller living space per inhabitant. Furthermore, private motorized transport is less attractive in cities than local public transport. Both lead to higher efficiency through sufficiency.

References

[BBVA 2012]   The future of world car fleet: The road ahead. BBVA research (2012)
[ENER 2011]   ENERCON erhält Zertifikat für Lebenszyklusanalyse, Windblatt 4:11 (2011)
[EROI 2013]   Weißbach, D., Ruprecht, G., Huke, A., Czerski, K., Gottlieb, S., Hussein, A.: Energy intensities, EROIs (energy return on invested), an energy payback times of electricity generating power plants. Energy **52**, 210–221 (2013)
[IEALCA 2015]   Life cycle inventories and life cycle assessment of photovoltaic systems. IEA International Energy Agency (2015)

[Kiel 2018]   Küstenkraftwerk K.I.E.L. Informationsbroschüre der Stadtwerke Kiel
   (2016)
[Mar 2002]   Marheineke, T.: Lebenszyklusanalyse fossiler, nuklearer und regenerativer
   Stromerzeugungstechniken (Juli 2002)
[Payk 2010]   Payk, B.: Entwicklung der Wohnflächenversorgung in Städten und
   Gemeinden Baden-Württembergs. Statistisches Monatsheft Baden-Württemberg **1**, 10–14
   (2010)
[Schaeff 1977]   Schäff, K.: Die Entwicklung zum heutigen Wärmekraftwerk. VGB
   Technische Vereinigung der Großkraftwerksbetreiber (1977)
[Siemens 2014]   Kraftwerk Lausward: Ein Kraftwerk – drei Weltrekorde. Siemens Power
   and Gas Division (2014)
[StatDE 2010]   Statistisches Bundesamt Deutschland, Statistisches Jahrbuch
[Stephan 2015]   Stephan, T.A.: Untersuchung der Wachstumsmuster zur langfristigen
   Prognose von Automobilmärkten in frühen Entwicklungsstadien. Dissertation an der
   Universität Duisburg-Essen (2015)
[UBM 2018]   Mobilität privater Haushalte. Verkehrsaufkommen im Personentransport.
   Umwelt-Bundesamt 16.05.2018
[VDE 2012]   Erneuerbare Energie braucht flexible Kraftwerke. Studie des VDE (2012)
[VES 2011]   Life Cycle Assessment of Electricity Production from a Vestas V112
   Turbine Wind Plant. PE NEW (2011)

Efficiency of Storage Technologies

5

5.1   Storage Technologies

Fossil energy supply from thermal power plants was carried out according to a power plant development plan in line with the grid load. Storage facilities were only necessary to a small extent. In particular, in the case of less flexible thermal power plants such as nuclear and lignite-fired power plants, generation profiles with minor changes in output were run. When the energy demand was low at night, the excess output of the power plants could be absorbed in pumped storage and fed back as a peak load at midday and evening of the following day. For this purpose, the pumped storage facilities were designed as short-term storage for storage times of a few hours.

Renewable energy supply from wind energy and photovoltaics has low full load hours with higher power per energy unit. Periods of several days with high overproduction can alternate with longer periods without sufficient supply (dark doldrums). The storage technologies must be designed for more flexible use with more frequent charge–discharge cycles and larger storage capacities, i.e., longer charge and discharge times.

The previously dominant central, large-scale pumped storage facilities are currently being converted for larger storage capacities and larger numbers of cycles in accordance with technical and environmentally relevant options.

New storage technologies are also emerging. In the field of decentralized energy supply, these are stationary accumulators in buildings with photovoltaics in order to increase the degree of utilization of the PV and to avoid high export power with grid overloads in the distribution grid. Mobile storage in electromobility supplement the share of decentralized storage capacities.

Compressed air storage as adiabatic or diabatic storage enables energy storage by means of compressed air in salt caverns. Currently, such a power plant exists in Europe in Huntorf [Huntorf 1980], [Huntorf 2001].

The sector coupling to the heating grids (power-to-heat) enables electric boilers to convert excess electrical energy into heat. The conversion of electricity into

© Springer Fachmedien Wiesbaden GmbH, part of Springer Nature 2022
G. Brauner, *System Efficiency by Renewable Electricity*,
https://doi.org/10.1007/978-3-658-35138-0_5

Table 5.1 Time ranges for the use of storage tanks in the energy supply

Storage type	Year	Month	Week	Day	Hour	Minute	Second
Pumped storage	x	x	x	x	x		
Hydrogen storage		x	x	x	x		
Compressed air storage		x	x	x	x		
Accumulator			x	x	x	x	
Steam storage					x	x	
Hot water tank			x	x	x	x	
Superconducting coil						x	x
Flywheel						x	x

hydrogen (power-to-gas) enables electromobility based on fuel cells with hydrogen produced from renewable sources.

Steam storage, also called root storage, can store superheated steam for steam power plants in the minute range. They are not very suitable for long-term storage.

Superconducting storage units and flywheel storage systems are used in particular in the area of sensitive industrial processes to intercept short-term voltage drops in the range of minutes and seconds in the event of grid disturbances. They do not play a role in the renewable energy supply.

Table 5.1 summarizes the storage types with their temporal areas of application.

5.2   Tasks of Storages

Storage can perform a variety of tasks in electrical energy supply. They can compensate for fluctuations in generation and store the balancing energy required for this purpose. Furthermore, they can participate in grid control in the area of frequency and voltage. They can also improve the slow gradient behavior of thermal power plants by temporarily intercepting the energy system with high power gradients in the event of grid disturbances until the power plants are able to replace the failed energy. Due to the changing generation technology with the replacement of a predominantly fossil energy supply by a predominantly renewable one, the necessary properties and capacities of the storage in the area of central and decentralized storage have to be reassessed and redefined. From today's perspective, pumped storage represents the essential *central* technology in the future, and accumulators in buildings with photovoltaics and in electric vehicles will be the most important *decentral* storage technologies. Both storage technologies can be used flexibly with low latency times when required and at high gradients. Some properties that stand out are listed below.

Store balancing energy
In the extra-high voltage transmission grid and in the high-voltage grid strong fluctuations in the generation are possible with a high proportion of wind power,

leading to surplus production and with lulls to insufficient supply. The loads in the grid can only follow these fluctuations to a limited extent and require a secure basic supply. By using storage systems, excess energy can be absorbed and fed back into the grid in periods of low generation. This increases the efficiency of renewable generation and reduces the need for fossil balancing energy.

Gradient leveling (gradient reduction)

With the expansion of renewable energy supply, generation gradients increase sharply. The residual power plants, which have to supply balancing energy in the event of gaps in production, must then be able to produce large power gradients. In thermal gas- and coal-fired power plants, high power gradients due to thermal stresses shorten the service life. Pumped storage power plants (PSPP) and accumulators are low temperature storage devices without heat stress problems. They are capable to performance large power gradients ranging from seconds to minutes (Fig. 5.1). They can initially apply these power gradients and relieve the slower thermal power plants. The thermal power plants can ramp up with their structurally-related lower gradient and the storage units are then ramped down accordingly. Coal-fired power plants in hot start (downtime less than eight hours) require two to three hours to restart. Combined cycle gas and steam power plants (CCPP) in the standard version take about an hour. First, the gas turbine starts up. A steam boiler is heated up with the hot exhaust gas until the steam condition is suitable for the operation of the subsequent steam turbine. This is shown by the horizontal line in Fig. 5.1. Flexible combined cycle power plants can reach their nominal output within half an hour [Tomschi 2008], [VDE 2012]. In this power plant from Siemens Power Generation, a steam storage device is charged, when it is shut down as a result of arising renewable generation. When restarting in hot start, the superheated stored steam is used to preheat the steam turbine immediately after the gas turbine has started. This makes it possible to avoid the waiting time (Fig. 5.1, CCPP flexible).

As Fig. 5.2 shows, the large gradient capability of pumped storage and accumulators can be used to relieve the thermal power plants of high gradients. First, the power is then provided with large gradients from pumped storage or accumulators

Fig. 5.1 Power gradients of storage facilities and thermal power plants

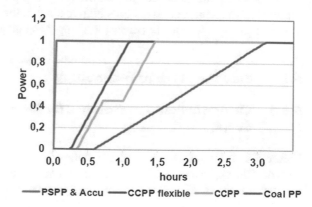

Fig. 5.2 Superseding of a fast-starting pumped storage facility by a combined cycle power plant (Power curve of the pumped storage power plant)

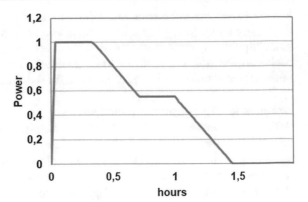

until the thermal power plants start up with their lower gradients. The pumped storage and accumulators then ramp down their power according to their limited storage capacity, and the thermal power plants can then provide balancing power for the energy system even for long periods. This interoperation of generation and storage combines the good properties of both types of plants at an economical and technically functional overall system.

Load Leveling (load shifting)
In renewable energy supply, the profiles of generation and load diverge. The generation is led by the renewable sources and the load characteristics are largely independent of it. To achieve the highest possible coverage rate of renewable energy for the load, renewable peak generation above the demand must be stored and shifted to periods with lower production at higher load demand.

Avoiding peak load (peak shaving)
In the industrial sector short-term load peaks can be covered by local accumulators. The investments required for this can be amortized by avoiding peak power dependent components of the grid tariff. Also, in the area of fast charging stations for electric vehicles, accumulators can enable high charging power in the range of 150 kW for about some hours and avoid thus high loading in the distribution grid. They can then be recharged with lower power for a longer charging period from the grid.

5.3   Pumped Storage Technology

5.3.1   Change in Pumped Storages for Regenerative Energy Supply

Figure 5.3 shows an overview of the pumped storage power plant in Kaprun in Austria. The upper reservoir Moserboden and the lower Wasserfallboden, each have a large volume of 80 million m³. Originally, since its commissioning in 1955, the

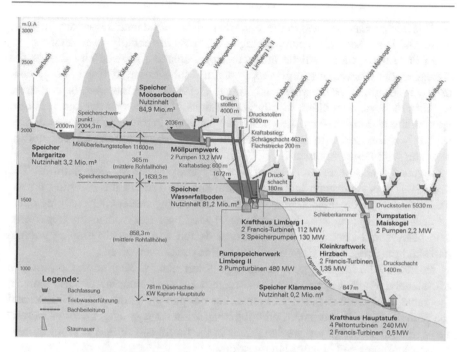

Fig. 5.3 Overview of the pumped storage power plant Kaprun with the upper stages Limberg I & II of Verbund Hydro Power (Verbund)

reservoirs was designed with pumps for a total of 112 MW and a maximum pumping time of 28 days for a complete circulation of the lower reservoir. In 2011 a new Limberg II upper stage was commissioned with a pumping and turbine power of 480 MW each [Limberg II 2007]. This reduces the pumping time to 165 h. Currently, another Limberg III upper stage is in preparation, which will change the plant to a total power of 960 MW with a pumping time of 80 h. The increase in power at sufficiently long pumping time or sufficiently large storage capacity is necessary due to the expansion of renewable energy, in particular of the wind energy.

Kaprun, with its symmetrical storage facilities with large volumes, is a special case in the field of Austrian pumped storages. Since the reservoirs were already in place, the approval procedures for the expansion were simplified, as only a new penstock with a larger cross-section and a new power plant cavern including pumps, turbines and generators had to be created. The expansion is not visible from the outside.

The upper reservoir is filled by precipitation and melt water directly and indirectly via the bypasses from neighboring valleys, about twice a year. The excess water can be processed via a main stage plant (Krafthaus Hauptstufe) and discharged into the river Kapruner Ache.

Many pumped storage power plants were designed as annual reservoir without pumping possibility. Over the year the rainfall and the water from snow melting was collected and have generated via turbines electricity.

This design was practical in the past when electricity demand was low. Some of these could also be used as pump storage with small lower daily reservoirs.

In the future, there will be an increasing demand for electricity due to the change to the renewable energy supply and the substitution of fossil energy end-use by renewable electricity. Annual storage facilities must therefore be converted to pumped storage facilities with larger volumes of the lower reservoir. It is also appropriate to convert an existing daily pumped store to a weekly pump store. This is mostly associated with the expansion or new construction of the lower storage reservoir with larger volumes.

Figure 5.4 shows an overview of the Austrian pumped storage facilities. The horizontal axis represents the storage time and the vertical the pumping power. The diameter of the circle is proportional to the energy content per pumping operation at full displacement of the volume of the lower reservoir to the upper one. The many small circles show a traditional design with limited pumping volume and short pumping time. Here future expansion options are given.

In case of the Limberg pumped storage power plant in Kaprun in the province of Salzburg, Fig. 5.4 shows that the storage energy per pumping process remains the same after the expansion measures, but the pumping power is increased and the pumping time is shortened. With this expansion strategy, larger surplus capacities can be absorbed at low market prices in the future and shifted to times of generation shortages and higher scarcity prices.

A second kind of capacity expansion is visible in the example of the pumped storage plant Kühtai in Tyrol. The power plant has the upper reservoir Finstertal with a volume of 60 million m^3 and a small lower reservoir Längental with only 3 million m^3 and a pump power of 242 MW. This results in a pumping time of only 12 h. This is a typical daily storage volume from the time of the hydrothermal generating scheme. A new lower reservoir at Kühtai with a storage capacity of 31 million m^3 and a pumping power of 180 MW results in a pumping time of 123 h or 5 days. This is a suitable expansion of the storage capacity for the regenerative energy supply.

In the future, in Austria and the neighboring countries in the Alps, mainly this second strategy of expansion of lower reservoir will be possible. Many annual

Fig. 5.4 Pumped storages in Austria, development trends

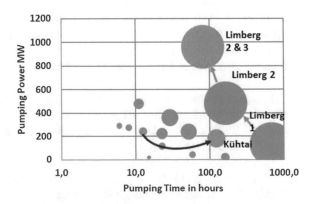

storage facilities in the Alps, which are only designed for turbine operation and not for pumping, need an adaption to the changed renewable energy production situation in Europe.

In total in Austria in 2016 pumped power of 3500 MW at an average cycle energy of all pumps of 227 GWh and an average storage time of 64 h is available.

5.3.2 Energy Economy of Pumped Storage in a Renewable Environment

Like all generation facilities, pumped storages are in the free electricity market. They fulfill important functions here:

- Storage of surplus energy, in the future especially renewable energy
- Feeding back energy into the grid in the event of a generation shortage at market prices
- Provision of balancing energy with high power gradients to relieve the slow thermal and nuclear power plants
- Provision of energy for frequency control
- Defense against major incidents through rapid start-up
- Network restoration due to a prepared black start capability

The construction of new pumped storage is connected with upper and lower reservoirs including the dams, as well as the penstock and the machine cavern. The motor-generators as well as the pumps, turbines, transformers, protection and control technology form the machine equipment. In today's high-capacity plant projects, the grid connection is made to the 380 kV transmission network.

The power of pumped storage is determined by the head between the reservoirs and the flow volume in m³/s. About 80% to 90% of the investment costs are for the dam, penstock, cavern and machine equipment. The expansion volumes of the reservoirs only account for around 10 to 20% of the investment costs. Pump storage facilities are therefore characterized in their specific investments according to a power price in €/kW. The investment costs are in the following ranges:

- Construction of new pumped storage facilities: 1000 to 2000 €/kW
- Conversion and expansion of existing pumped storages: 500 to 1000 €/kW

For accumulators, the investment costs are not determined by power, but by the storage capacity in €/kWh, as will be shown later.

Energy economy of new pumped storages
Pump storages have a long service life. The dam and penstock have a life time of 80 to 100 years. The machine equipment can be used for around 50 years. For economic reasons, pump storage facilities will be depreciated in around 20 years. New pump storage facilities have high depreciation rates in a period much lower

than the service life and are therefore uneconomical within this period. The profitability of the pump storage results from its operation as a pump storage pool, in which the already depreciated facilities are valued together with the new plants at a mixed price.

The following prerequisites are used for the energy industry analysis:

- Interest rate on the capital employed: $i = 3\%$
- Depreciation period: $n = 20$ a
- Market price of the pump energy: $c = 0$ to 3 /kWh
- Operation and maintenance costs: $om = 20$ €/kW/a

The market price for the electrical energy supplied by pumped storage (generation costs) is determined by the annual number of operating hours. Up to about 4000 h are possible in pump mode and just as many in turbine mode.

Figure 5.5 shows the generation costs of completely newly built pump storage plants over the full load hours in pump or turbine operation. The maximum generation costs apply to the upper specific investment costs and a market price for pump energy of 3 ct/kWh, the minimum generation costs apply to the lower specific investment costs and a price of 0 ct/kWh for pumping energy.

The generation costs decrease with high full load hours. This also means that if there is overcapacity in pumped storage facilities, the generation costs rise.

With 2000 full pump operating hours, the costs are between 7 and 13 ct/kWh. Individual, newly built pump storage facilities are therefore not economically viable on the market during the depreciation period.

The conversion or expansion of existing pumped storage facilities, as is currently the case with most new construction projects, results in lower generation costs (Fig. 5.6).

The generation costs with conversions or expanding of existing pumped storage facilities at 2000 full load hours are in the range of 5 to 9 ct/kWh. With minimal

Fig. 5.5 Generation costs of new pumped storage buildings

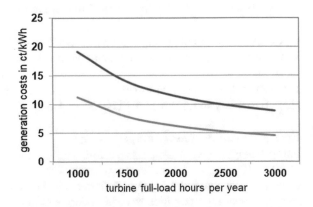

Fig. 5.6 Generation costs when converting or expanding pumped storage facilities

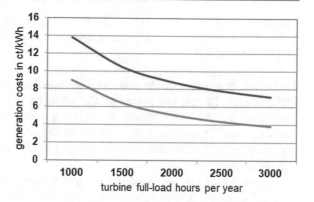

expansion costs and high full load hours of 3000 h/a, they reach 4 ct/kWh and are thus in the range of the future generation costs of wind energy. With this high temporal utilization of the pump storage, the stored energy is as expensive as that generated in renewable plants. Pump storage systems should therefore be counted as renewable generation plant in the future and continue to be relieved of all additional costs in order to be able to offer inexpensive energy.

If, on the other hand, individual plants that have not yet been depreciated go on the electricity market at individual generation costs, they have no chance against cheap energy from fossil or nuclear power plants. This would mean that surplus renewable energy has to be shut down and replaced by fossil or nuclear energy. The energy prices on the exchanges for base and peak are around 3 to 7 ct/kWh. At least in the peak area, pump storage with low investment costs (Fig. 5.6) are economical.

Energy economy of pool systems

In future, the expansion projects for pumped storage in Austria will show, that about one-third of the existing pumped storage plants are in the depreciation period and that two-thirds have low generation costs as depreciated plants. The lower curve in Fig. 5.6 can be considered as the mean cost curve for all pumped storages, operated as pools. This means, that all pump storage facilities are economically viable in the peak price range on the market.

Due to the decommissioning of nuclear and coal-fired power plants in Germany and increasing bottlenecks in the European transmission grid, it is expected that periods with higher peak prices will increase. This will further improve the profitability of pump storage in the future. New regulatory measures that give priority to stored renewable energy over fossil balancing energy can further improve the situation of pumped storage and thus the sustainability of the electricity supply.

In Austria, all pumped storage facilities have to pay a network usage charge of 0.233 ct/kWh for work and 100 ct/kW for power, regardless of the network level they feed into [SNT 2017]. At 2000 full load hours, this means a load of 0.283 ct/ kWh. In Switzerland and Germany, pump storage is exempt from this [stoRE 2014], [Auer 2014].

Comparison of pumped storages and accumulators
For pumped storage, the power of the plant essentially determines the specific investment in €/kW. With accumulators, the energy content determines the volume and therefore the specific investment in €/kWh. Pump storage is a central facility and have the task of storing renewable energy and providing balancing energy for the entire energy system.

Accumulators are decentralized storage devices. Their task is to store locally renewable energy and to reduce the grid load on the distribution grids and thus to reduce the grid expansion to an economically justifiable level. In the case of accumulators, the specific capacity costs in €/kWh are therefore largely independent of the total storage capacity. In the case of pumped storage, on the other hand, they decrease with the storage capacity. Pumped storages allow a great number of charging and discharging cycles. With accumulators the depth of discharge and the amplitude of charging and discharging current limit the service life.

Figure 5.7 shows the specific capacity costs (investment costs) for Li-Ion accumulators with an estimate of the expected reduction in costs through large-scale production by 2050.

Pump storage with its specific investments of 1000 to 2000 €/kW for new plants and 500 to 1000 €/kW for conversions or expansions will also be competitive in future in terms of their capacity costs compared to Li-ion accumulators, if they have storage capacities in relation to their nominal power for at least 15 to 20 h.

The construction and conversion strategy for pumped storage is moving in the direction of expanding capacity as a weekly storage, which will keep it competitive with accumulators in the long term. Pumped storages and accumulators are only partially in competition with each other, since, as has already shown, they have different areas of application.

Fig. 5.7 Comparison of the specific costs of pumped storage and accumulators

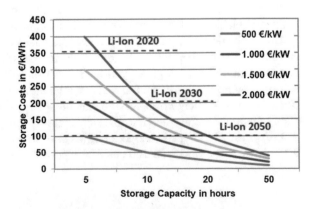

5.4   Compressed Air Storage

In the lowlands, the construction of compressed air storage power plants is an alternative to pumped storage power plants. In these power plants, air is compressed by compression turbine, stored in underground caverns and, when expanded, electrical energy is recovered in turbines.

The following types of compressed air storage power plants are possible:

Diabatic Compressed Air Energy Storage plant, D-CAES
Figure 5.8 shows the principles of the diabatic and the adiabatic compressed air energy storage power plants. Both have a motor-generator (MG) that can either drive the compression turbine shown on the left as a motor or is driven as a generator by the expansion turbine shown on the right.

With the diabatic compressed air energy storage power plant (D-CAES) the air is cooled during compression in the compressor and in the downstream heat exchanger and the compression heat is lost. With the expansion in an air turbine, the air cools down and the outlet temperature can be very low. It is therefore necessary for heating the compressed air supply to counteract icing. Diabatic compressed air power plants have been just as hybrid diabatic versions realized, in which natural gas is mixed with the compressed air and the hot exhaust gas drives a gas turbine. Their efficiency is low, around 40 to 50%.

Hybrid Diabatic Compressed Air Energy Storage with gas turbine, HD-CAES
With this design (HD-CAES), the air is compressed in the storage mode as described above and cooled to the permissible temperature of 50 °C in the storage

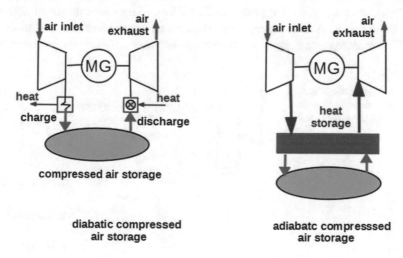

Fig. 5.8 Diabatic and adiabatic air compressed air storage power plant

cavern. The compression heat is given off to the environment and is therefore lost for the expansion mode. In the turbine mode, the compressed air is mixed with gas and directly fed into a gas turbine. By this a compressor can be omitted, which at least needs 60% of the total power of a normal gas turbine. The air is mixed with natural gas and burned in the combustion chamber and is expanded in the gas turbine to atmospheric pressure. Such an air storage gas turbine power plant (Fig. 5.9) was put into operation in Huntorf in Northern Germany with two salt caverns as storage facilities in 1979 [Huntorf 2001]. The power plant has a compressor power of 60 MW and requires eight hours to charge the two caverns [Weber 1975]. The discharge takes place in two hours via the gas turbine with a generator power of 290 MW. The pressure in the two caverns is between 46 and 66 bar [Huntorf 1980]. The Huntorf HD-CAES is a peak load power plant. The starting time from standstill is 11 min [Huntorf 2001].

In Huntorf there are two cylindrical caverns of 135,000 m³ each under a mountain top layer of 500 m. The caverns have a diameter of 30 m and a height of 200 m. They were washed out by a brine process with fresh water. The salt load could not be discharged into rivers but was discharged into the brackish water of the Elbe estuary via a pipeline 40 km in length. In inland areas far from coasts, the brine process poses an environmental impact assessment problem, since the European water framework directive will not allow introducing brine into rivers.

In 1991, a hybrid compressed air storage power plant with an output power of 110 MW was put into operation in McIntosh, Alabama.

In 1986, a demonstration plant with compressed air storage in porous sandstone was temporarily put into operation in Sesta in Italy but shut down because of lack of storage capacity in the porous sandstone.

An HD-CAES plant with an air pressure storage in porous sandstone was planned in 2015 in the USA. The Iowa Stored Energy Plant [ISEP 2012] project was to achieve a compressor power of 220 MW and a turbine power of 270 MW with 36 full-load turbine hours. The project was also abandoned due to of the insufficient storage capacity of the porous sandstone.

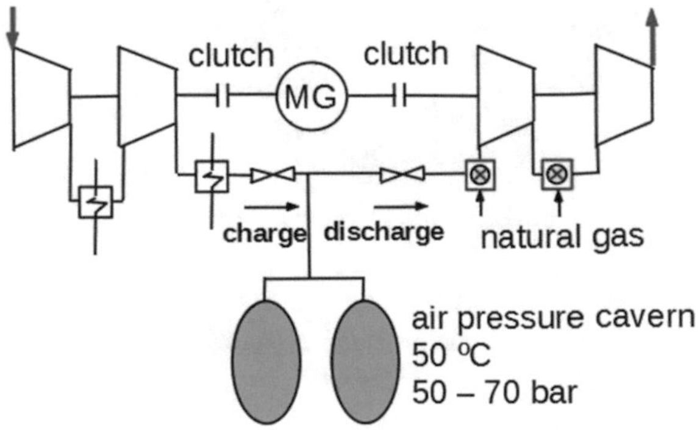

Fig. 5.9 Schematic of the Huntorf air pressure storage power plant [Huntorf 1980]

Fig. 5.10 Schematic of an
adiabatic compressed air
storage power plant [ADELE
2012]

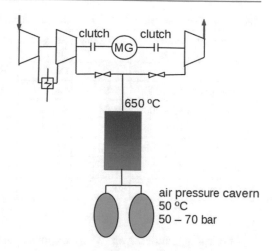

650 °C

air pressure cavern
50 °C
50 – 70 bar

Adiabatic Compressed Air Energy Storage power plant, A-CAES

In the adiabatic compressed air energy storage, the is air not totally cooled during compression. Since very high temperatures of up to about 900 °C are possible with single-stage compression, two-stage compression is usually used. After the first compression stage, the hot air is cooled without storage in order not to exceed the planned temperature of the heat storage of about 650 °C after the second stage (Fig. 5.10). After passing through the heat exchanger unit, the air is cooled down to the allowable temperature of the salt cavern of 50 °C.

During the expansion process, the air from the compressed air caverns first passes through the heat exchanger and is then expanded to normal pressure and ambient temperature in the air turbine on the right in the picture. The heat storage is only suitable for short-term operation and would discharge itself through heat losses, if there were longer periods between compression and expansion.

RWE investigated the feasibility of an adiabatic compressed air energy storage for electricity supply in project ADELE. The target data were 200 MW compressor power, 250 MW turbine power, a storage capacity of 1 to 2 GWh and 4 to 8 turbine full-load hours [ADELE 2012], [ADELE 2013]. A salt cavern was to serve as storage. The efficiency should be 70%. The project was discontinued in 2015.

5.5   Comparison of Storage Technologies

A-CAES compressed air storages use the thermodynamic change in state of the air. The storable energy per cycle depends on the volume of the compressed air storage, the upper and lower pressure of the storage and the upper temperature of the heat storage.

Table 5.2 Characteristic values of new storage technologies

Storage type	Power range	Efficiency	Start-up time	Max gradient	Investment costs	Service life
	MW	%	min	%/min	€/kW	Years
Pump storage	10–1000	75–85	1–2	100	500–2000	50–100
D-CAES	100–400	40–54	9–15	20	600–1500	40–50
HD-CAES	100–400	–	20–35	5	900–2000	30–40
A-CAES	100–400	60–70	15	10	700–1700	40–50
Li-Ion battery	1	80	<1	>100	400–1000 €/kWh	6–10
lead battery	1	80	<1	>100	100–200 €/kWh	5–8

The energy and power of hydropower pumped storage per working cycle depends on the head between the reservoirs, the volume of the (smaller) lower reservoir and the flow rate of the pumps or turbines.

Pumped storage can do a much larger cycle work than compressed air storage. The cycle work of the Limberg II pumped storage plant is 120 times greater than the energy content of the two caverns of the Huntorf compressed air storage facility. For the same cycle capacity, 240 salt caverns would have to be created and a large number of compressed air storage power plants built. Compressed air storage can be used in particular as a smaller decentralized storage facility.

Table 5.2 shows a comparison of the storage technologies. The investment costs for compressed air storage systems are now in the area of pumped storage power plants. In the future, the adiabatic compressed air storage will be of particular interest because it does not use natural gas and is therefore emission-free. Furthermore, they operate at efficiencies of 70%. Compared to pumped storages and accumulators, compressed air storage systems show less favorable values for efficiency, start-up times and maximum power gradients.

Pumped storage is the most economical storage facility in terms of technology and energy. They have a long service life, which extends to 100 years for structural systems and 50 years for machine systems. After depreciation, they have only low operating costs. Pumped storage will therefore be the key technology for large central storage in the future. As a decentralized storage technology, accumulators with lower power and capacity will be important in the future, and also have favorable properties in the areas of efficiency and start-up time, power gradient.

5.6   Accumulators

Comparing technologies of accumulators

Accumulators are secondary cells. With these, electrical energy is absorbed during the charging process and released during the discharging process. In contrast, batteries are primary cells. They decompose when discharging and cannot be recharged. Unfortunately, in power engineering, the term "storage battery" instead of "accumulator" is often used. Only the term "accumulator" should be used correctly. There are many research and development activities in the field of accumulators, particularly for electric vehicles, stationary energy storage devices and mobile devices from the field of information and communication technology.

The characteristics of accumulators depend on external factors such as depth of discharge, ambient temperature, discharge current in relation to the capacity and the number of charging cycles. The *State of Charge SOC* describes the state of charge as a percentage of the nominal capacity

$$SOC = \frac{available\,charge}{nominal\,charge} \cdot 100\ [\%\,]$$

The *Depth of Discharge DOC* is

$$DOC = 100 - SOC\ [\%\,]$$

The *Capacity Factor C* represents the ratio of the charging or discharging current to the nominal current. The nominal current is defined for a discharge within one hour.

Example:

A lithium-ion accumulator has a capacity of 20 Ah and is discharged at 4 C. The nominal current is 20 A. The value 4 C corresponds to a discharge current of 80 A.

To maintain their capacity, accumulators must neither be overcharged nor deeply discharged. The end-of-charge voltage is the upper voltage limit that chargers must comply with during the charging process. The deep discharge voltage represents the lower permissible voltage during discharge. Figure 5.11 shows the course of the voltage of a lithium-ion accumulator with different C-factors.

The charging voltage of 3.6 V changes to the nominal voltage of 3.3 V at lower capacity factor ($C \leq 1$). The low internal resistance of only about 10 mΩ of this LiFePO$_4$ cell of 2.3 Ah from manufacturer A123 means, that the cell capacity limits at a discharge voltage of 2.0 V are very close to one another, even with high discharge currents.

Fig. 5.11 Course of the discharge voltage of LiFePO$_4$ accumulators on SOC at various C factors [A123 2010]

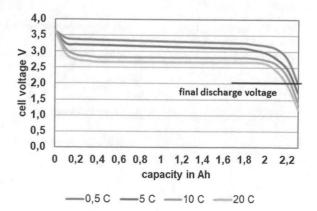

Accumulators age considerably at increased operating temperatures and high discharge currents, as shown in Fig. 5.12. The utilization limit is defined for a remaining capacity of 70%.

With a discharge at 1 C and an ambient temperature of 25 °C, more than 6000 cycles are achieved. With a discharge with 2 C at 60 °C there are only 1500 cycles. High temperatures of accumulators with poor cooling and frequent rapid charging shorten their service life. Liquid-cooled accumulators within narrow temperature limits between 5 °C and 25 °C will have a long service life.

The effect of the depth of discharge (DOD) on the service life is shown in Fig. 5.13. Small depths of discharge are beneficial for a long service life. This also means that the capacity of accumulators should be not too tight designed as then frequent deep discharges shortened the life time too much.

Currently, the following **accumulator types** are common.

Lead-acid accumulator This is the classic accumulator with low investment costs. Because of its low energy density, it has been replaced by lithium-ion technology as a drive storage device in electromobility. It is still used in the field of stationary storage.

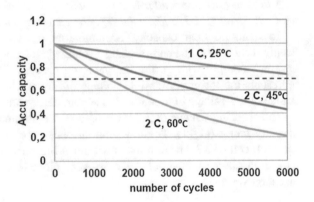

Fig. 5.12 Number of cycles of a LiFePO$_4$ accumulator

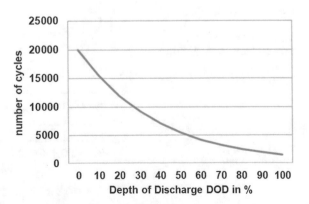

Fig. 5.13 Influence of the depth of discharge on the number of cycles

Nickel–cadmium accumulator (NiCd) This storage was of great importance and was the mostly used technology until 1990. The Lithium-ion accumulator has this type replaced, due to its higher power density and lower self-discharge. Because of its cadmium content, the NiCd accumulator has been banned in the EU since 2004 and may only be used for special applications such as replacements in power tools and in medical technology.

Nickel–Metal hydride accumulator (NiMh) This type of accumulator is very common in the form of round cells (AA and AAA).

Nickel–Metal hydride with Low Self-Discharge (NiMh-LSD) Since 1990, this type of low self-discharge of 1 to 2% per month in the area of small accumulator applications in use. In large capacity installations, he has so long not been used.

Lithium Cobalt dioxide accumulator ($LiCo_2O_2$) This is the most common Li-Ion accumulator, having a cell voltage of 3.7 V. Lithium-ion accumulators have a very low cell internal resistance. As a result, the heat development in the accumulator during charging is low and rapid charging is possible. When discharging, high currents in the range up to about 20 C are possible for a short time. Because of its high energy density of up to 140 Wh/kg, it is widely used in electromobility.

Lithium iron phosphate accumulator ($LiFePO_4$) This accumulator is becoming more and more popular in the field of electromobility. It is superior compared to lithium cobalt dioxide accumulator by low fire risk, fast charging ability and low self-discharge. It has a slightly lower cell voltage of 3.3 V and a slightly lower energy density.

Sodium–sulfur high-temperature storage battery (NaS) This battery was used from 1980 to 1990 in the first generation of electric cars. It is characterized by a very high energy density of up to 220 Wh/kg, has an operational temperature range of 270 to 350 °C and an operating temperature of 300 °C. Although the accumulator does not have an electrochemical self-discharge, it requires electrical heating to maintain the operating temperature. This results in considerable self-discharge despite thermal insulation and the accumulator is no longer operational after a few days. The heating process up to operating temperature then takes several days. Vehicles of the pioneering generation of electromobility therefore had to be able to provide the heating energy when parking, at a charging station. In the event of accidents involving sodium-sulfur batteries, fires with high temperatures are possible. This led to the discontinuation of the use of this accumulator in the test phase with electric vehicles. This accumulator is still used internationally in stationary applications in energy technology.

Sodium Nickel–Chloride high temperature accumulator (NaNiCl) (ZEBRA-accumulator: Zero Emission Battery Research Activities). This high-temperature storage accumulator [Boehm 2010] was the second attempt on electric mobility from the year 2000, planned in electric vehicles such as the Mercedes A-class, and realized in Think City, Smart EV, Renault Twingo and Panda Elettrica [ATZ 1997]. With a storage capacity of 28.2 kWh in the car Think City, the heat loss and thus the average heating power from the accumulator itself was approximately 100 W. From this a fully loaded accumulator was discharged after 11 days, which corresponds to a self-discharge rate of 260% per Month.

In the meantime, the Li-ion battery has established itself in the electrical energy supply and in electromobility.

The developments in the accumulator sector are currently in a state of flux and further types with high rapid charging capability, low self-discharge, high energy density and, in particular, a long service life, are to be expected. The service life of the batteries is 6 to 10 years with normal use, or 20 years with intensive cooling. There are still research activities and developments to be expected.

Table 5.3 shows a compilation of characteristic values for the various accumulator technologies.

Currently, the lithium iron phosphate accumulator has great growth potential in electromobility and energy technology, particularly because of its high C-factors in high-current discharge and fast charging.

5.7   Load Leveling with Storage Capacities

With renewable self-generation, the load is shifted in time according to the renewable supply. In case of electromobility, this is possible simply by shifting the charging time. Similarly, electric heating systems with heat storage can also be charged at different times. If there is a divergence between demand and generation, the electrical energy can be shifted to the periods with demand via electrical storage. The profitability of the systems results from the difference between electricity purchased from the grid and the costs of the in-house generation system including storage unit.

Energy economic analysis
With variable electricity tariffs, it is possible to shift the peak load from high-tariff times to low-tariff times by using storage systems. With constant tariffs, it is particularly interesting to make the share of self-use as large as possible in the decentralized use of PV. This case will be investigated. Figure 5.14 shows the principle of load shifting.

The use of local storage increases the degree of self-using k_m compared to usage k_0 without storage. To determine the possible degree of utilization of photovoltaics with and without storage, a time series analysis based on measurements in Lower Austria is carried out.

The total energy generated annually with PV results from the installed PV power P_{PV} and the annual full load hours T_{PV}. The cost savings result from the energy difference between the degree of self-use with storage k_m and without

Table 5.3 Comparison of accumulator technologies

Technology	Cell voltage			Charging current		Charging cycles	Energy density	Self-discharge
	Nominal	Loading	Discharging	Nominal	Max	50% DOD		
	V	V	V				Wh/kg	%/month
Lead-acid	2.10	2.40	1.75–1.87	0.1 C	0.4 C	500	30–40	10–20
NiCd	1.20	1.45	0.85–1.00	1.0 C	2.0 C	650	40–60	10–20
NiMh	1.20	1.45	1.00–1.10	1.0 C	2.0 C	1000	70–80	15–25
NiMh-LSD	1.20	1.45	1.00–1.10	1.0 C	2.0 C	1000	70–80	1–2
$LiCo_2O_2$	3.70	4.20	2.50–3.00	1.0 C	3.0 C	3000	130–140	2–4
$LiFePO_4$	3.30	3.60	2.00–2.50	4.0 C	10 C	5000	90–120	2–5
NaS 300 °C	2.00	2.08	1.78–1.90	0.1 C	1 C	8000	120–200	280
NaNiCl 290 °C	2.58	3.05	1.58–2.00	0.1 C	1 C	5000	100–120	260

Fig. 5.14 Principle of load shifting with PV

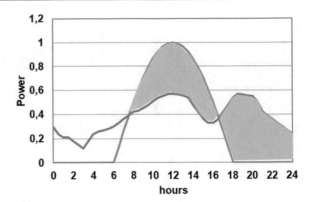

storage k_0 and the associated savings based on the avoided grid tariff c_n. The annual cost savings K_p through the use of local stores is:

$$K_p = P_{PV}T_{PV}(k_m - k_0)c_n \tag{5.1}$$

Without accumulator, the unused PV energy is fed back into the grid and is reimbursed by aggregators with a tariff c_R. The annual reimbursement K_R would be:

$$K_R = P_{PV}T_{PV}(k_m - k_0)c_R \tag{5.2}$$

The energy fed back tariff is without grid tariff and thus of lower. The investment in a storage system is economical, the annual depreciation on the storage system $\alpha \cdot K_S$ is less than the difference between the tariff savings and the potential reimbursement. The investment value of the storage K_S can be determined via the annuity α, which represents the annual portion of the repayment.

$$K_S \leq \frac{K_P + K_R}{\alpha} = \frac{(K_P - K_R)(q^n - 1)}{(q - 1)q^n} \tag{5.3}$$

With the annuity factor α, n means the usage time in years ($n = 10$) and q is the interest factor, with 3% interest the interest factor is 1.03.

As Table 5.4 shows, if the price of the accumulator of 4 kWh is not higher than 183 €/kWh so the storage is economical. The price range for storage systems for photovoltaics is between 250 and 1000 €/kW (2018). It is to be expected that through series production the prices will go in the direction of 100 to 200 €/kWh within a decade and that storage will become economical.

With a PV system of 2 kW, a storage capacity of 4 kWh improves self-use from $k_0 = 35\%$ to $k_m = 47\%$. An increase in the storage capacity from 4 to 5 kWh brings only a slight improvement in the coverage rate from $k_m = 47\%$ to $k_m = 49\%$ and that from 6 kWh to $k_m = 51\%$. Larger storage facilities are therefore not economical. They can only partially compensate the gap between the generation and load characteristics. The ratio of PV storage capacity to PV power is economical up to around 2 kWh/kW. This is discussed in Chap. 7 and was examined in detail based on time series analyzes of the PV supply.

Table 5.4 Profitability calculation for load shifting

Parameter	Dimension	Value
PV power P_{PV}	kW	2.0
Full load hours T_{PV}	h/a	1272
Storage capacity	kWh	4
Life time n	years	10
Interest rate	%	3
Annuity factor		0.1172
Self-use without accumulator	%	35
Self-use with accumulator	%	47
Network tariff c_n	€/kWh	0.30
Remuneration for feedback c_T	€/kWh	0.02
Network tariff cost savings	€/a	91.58
Potential reimbursement	€/a	6.1
Permissible accumulator investment price	€	730
Specific permissible investment price	€/kWh	183

5.8 Peak Load Avoidance (Peak Shaving)

This means avoiding high load peaks in industrial and commercial power supply. This is interesting because of the tariffs of this end consumer group, when the highest quarter-hourly peak load within a year is used for the peak load component of the reference tariff. The peak load can be reduced either by energy management with throttling of production or by feeding in additional power from a storage facility (Fig. 5.15). With renewable generation, storage capacities can also help to avoid feeding of peak power into the grid.

Fig. 5.15 Principle of peak load avoidance through storage use

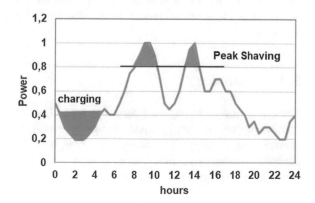

Energy Economic Analysis
A peak load can be reduced by using storage or load management. In the field of industry and commerce, the power share in the network usage fee is calculated based on the annual peak load and the work share based on the annual energy consumption. In addition, a network loss fee is determined based on the work share. From the point of view of the energy industry, measures to reduce the peak load are economical if the annual savings in the power component of the network usage fee are greater than the annual costs of the measures installed for this purpose.

This is to be investigated for an industrial company that is connected to the medium-voltage network.

Annual energy consumption E 30 million kWh
Peak load P 6000 kW
annual full load hours T 5000 h/a $(T = E/P)$
grid power price c_P 60 €/kW
grid energy and loss price MV   1.1 cent/kWh

Annual costs
    grid power price = 6000 kW * € 60/kW = 360,000 €/a
    grid energy price = 30 million kWh * 1.1 c/kWh = 330,000 €/a
    total grid costs = 590,000 €/a
    cost share of power per 1 kW = € 60/kW
    target: Lowering the peak load by 1000 kW

The costs of the storage system to be installed are determined by the duration of the power peaks and thus the storage energy required. When the power peaks occur rarely and only each takes a quarter of an hour, a relatively small storage of about ½ MWh can be used, which, however, must be quickly discharged and recharged. Thanks to the storage system, annual network service costs of € 60,000 can be avoided. With a service life of the storage of six years and an interest rate of 3% for the capital employed, the annuity factor is 0.1846. This results in an allowable investment price of 60,000/0.1846 = 325,000 € or of 650 €/kWh, so that the use of the storage for the peak load is economical.

In industrial companies, measures to avoid peak loads through load management are usually more economical. For example, by staggering industrial processes with high power requirements, the peak load can be reduced with less investment.

5.9 Summary

The expansion of storage capacities is necessary when the energy supply is predominantly renewable. Pumped storage is the most suitable technology for large central storage. Up to now many were designed either as annual storage without pumping option or as pump storage with small lower reservoir for only several hours of pumping operation. New construction or conversion to pump storage with storage capacities for about a week of rolling operation is appropriate. When marketing pumped storage capacities as a pool of new and depreciated systems, pumped storage will in future also be economical compared to decentralized accumulators.

In terms of accumulators, lithium-ion storage systems are gaining acceptance both as decentralized stationary storage system for photovoltaics and as mobile storage system in electric vehicles. PV storage should be designed as day storage. Larger capacities are not economical.

Compressed air storage should be designed as adiabatic storage. Compared to pumped storage, they have small capacities and the extraction of salt caverns is an obstacle in inland approval procedures.

References

[ADELE 2012]   Moser, P.: Druckluftspeicher. Vortrag auf dem Technik-Dialog 2012 der Bundesnetzagentur „Speichertechnologien". Bonn 2012. RWE Power AG

[ADELE 2013]   ADELE—Der adiabate Druckluftspeicher für die Elektrizitätsversorgung. Pressemitteilung von RWE (2014)

[ATZ 1997]   Die neue ZEBRA-Batterie auf dem Prüfstand. Automobiltechnische Zeitschrift 99, 5 (1997)

[Auer 2014]   Auer, H.: Hindernisse und Erfordernisse für Pumpspeicher in Österreich. www.store-projekt.eu

[A123-2010]   High Power Lithium Ion ANR26650m1A Data sheet. A123 Systems Inc. (2010)

[Böhm 2010]   Böhm, H.: Eine neue Hochleistungsbatterie. Technische Rundschau. Ausgabe 45(1990)

[Tomschi 2008]   Tomschi, U., Brauner, G.: Regel- und Ausgleichsenergie großer Windparks mit Gasturbinen. 9. GMA-Tagung Netz- und Systemführung, 5. und 6. März 2008, München

[Huntorf 1980]   Hoffeins, H., Romeyke, N., Hebel, D., Sütterlin, F.: Die Inbetriebnahme der ersten Luftspeicher-Gasturbinengruppe. Brown Boveri Mitteilungen 67(3), 465–473 (1980)

[Huntorf 2001]   Mohmeyer, K.-U., Scharf, R.: Huntorf CAES: More than 20 Years of Successful Operation. Spring 2001 Meeting, Orlando Florida, USA, 15–18 April 2001

[ISEP 2012]   Schulte, R.: Lessons from Iowa: Development of a Bulk Energy Storage Facility in the Midwest Independent System Operator (MISO) Market. 9 February 2012

[Limberg II 2007]   Wagner, E., Stering, P., Binder, E., Wimmer, K.: Pumpspeicherwerk Limberg II. Oesterr. Wasser Abfallwirtsch. 59(5–6), 69–70 (2007)

[SNT 2017]   398. Verordnung der Regulierungskommission der E-Control, mit der die Entgelte für die Systemnutzung bestimmt werden. (Systemnutzungsentgelte-Verordnung 2018-SNE-V-2018)

[stoRE 2014]   Facilitating energy storage to allow high penetration of intermittent renewable Energy. stoRE—Final Publishable Report (2014)

[VDE 2012]   Erneuerbare Energie braucht Flexible Kraftwerke. VDE-Studie (2012)

[Weber 1975]   Weber, O.: Das Luftspeicher-Gasturbinenkraftwerk Huntorf. Brown Boveri Mitteilungen 7(8), 332–337 (1975)

Efficiency of Grids

6

6.1 Tasks of the Electricity Networks

The electrical networks are used to transmit and distribute electricity. They consist of the following components:

- overhead lines
- cables
- transformers
- switchgears

Additional components are available for protection in the event of malfunctions and for control and monitoring:

- Selective protection devices
- Automation systems for control and monitoring

In Europe, in electrical energy supply the following voltage levels are used, which in the energy regulation are designated with the numbers 1 to 7 (Table 6.1).

In Europe, no voltage level higher than 380 kV has been built so far because of the dense settlement. The expansion of regenerative energy sources for a predominantly sustainable energy supply by the year 2050 has a major impact on the future design of the electricity networks. So far, the renewable energy plants have been integrated into the existing grids without significant network expansion. To this, the existing grid capacity reserve was released. The grids were previously characterized by a static safety philosophy. Here, the network capacities of the redundant transmission networks were only allowed to be used up to around 60%, so that in the event of a fault in a parallel line system, the remaining system could take over the full network load without further intervention by the system operator.

© Springer Fachmedien Wiesbaden GmbH, part of Springer Nature 2022
G. Brauner, *System Efficiency by Renewable Electricity*,
https://doi.org/10.1007/978-3-658-35138-0_6

Table 6.1 Voltage level according to European energy regulation

Level	Nominal voltage	Designation	Tasks
1	380/220 kV	Transmission grid, European interconnected grid, extra-high voltage grid	Long-range energy transmission, integration of large-scale power plants and wind energy parks
2	380–220/110 kV	Main transformers to the high-voltage grid	Coupling of network levels 1 and 3
3	110 kV	High-voltage grid of the regional utilities	Local grids approx. 300 MVA each Integration of large power plants and wind energy
4	110/10–20–30 kV	Transformers to the medium voltage network	Coupling of network levels 3 and 5
5	10–20–30 kV	Medium voltage grid	Local grids approx. 30 MVA each Integration small power plants with CHP and individual wind turbines
6	10–20–30/0.4 kV	Transformers to the low voltage network	Coupling of levels 5 and 7
7	0.4 kV	Low voltage grid	Local grids approx. 1 MVA each Integration of photovoltaics and mini power plants with CHP

This principle of static security has been replaced by dynamic security. Here, the lines can now be used up to the load limit. If the permissible network loads are exceeded, congestion management is carried out. Here, the system operator can intervene in the energy market and change the feed-in power of power plants or wind farms, that cause congestions. Furthermore, power plants can be started in remote regions, if this eliminates bottlenecks. Congestion management has increased significantly due to insufficient expansion of the grid compared to the expansion of wind energy. In Europe, wind farms in the north of Central Europe are often shut down and thermal power plants in the south are started up instead. The energy that is not fed into the grid is remunerated for the wind farms, and the energy fed into the grid is paid to thermal power plants.

Large wind farms are integrated into the 380 kV transmission grid only. Smaller wind farms up to about 200 MW can be integrated into the 110 kV grids. In the medium-voltage level only individual wind turbines as a village wind plants for the decentralized supply of settlements are found.

Photovoltaic systems are possible as central large-scale systems in the range of up to 100 MW in open-field installations or as decentralized small-scale systems on buildings. In the future, the funding strategies will tend to focus on small systems integrated into the building:

- There is no use of the landscape and the land area is available for the production of biomass or for reforestation.

- The solar electricity generated can at demand-based planning directly be used or stored to a large extend in the building and therefore requires no expansion of the low-voltage grid.
- Central large-scale PV systems, on the other hand, require grid connections according to the installed system capacity. However, the peak power occurs only some 100 h per year, which is an inefficient grid installation.
- In the future free electricity market, PV must offer their energy on the electricity exchange. In large weather situations with high global radiation, wide area over-production occurs and market prices fall to the marginal cost, which are near to zero for PV systems. This does not occur with decentralized use in building integrated power plants.

If designed appropriately with local storage systems, PV systems on buildings can deliver all of their energy to the end user without overloading the distribution grids with significant energy exports. When owned by the end users, the systems are also economical, because the energy costs of PV are comparatively lower than the grid tariff.

6.2 Theory of Line Transmission

The transmission of electrical energy via overhead lines and cables is described by line equations. They represent the relationship between the voltages U_1 at the input and U_2 at the output and the currents I_1 and I_2.

$$U_1 = U_2 cosh\gamma l + I_2 Z_W sinh\gamma l \qquad (6.1)$$

$$I_1 = \frac{U_2}{Z_W}sinh\gamma l + I_2 cosh\gamma l \qquad (6.2)$$

Here Z_W represents the operational characteristic impedance of the line and γ the propagation constant. These can be calculated from the line parameters:

- Inductance L' in Henry per km
- Capacitance C' in Farad per km
- Resistance R' in Ohm per km
- Leakage G' in 1/Ohm per km
- Circular frequency $\omega = 2\,\pi f\,(f = 50\ \text{Hz})$

Propagation constant of the line:

$$\gamma = \alpha + j\beta = \sqrt{(R' + j\omega L')(G' + j\omega C')} \qquad (6.3)$$

Here, α denotes the attenuation per km of propagation on the line and β the phase angle in rad per km. The operational characteristic impedance represents the relation between the voltage and current in propagation along the line.

Operational characteristic impedance of the line:

$$Z_W = \frac{(R' + j\omega L')}{\gamma} = \sqrt{\frac{R' + j\omega L'}{G' + j\omega C'}} \tag{6.4}$$

The operational characteristic impedance applies to the symmetrical operation of overhead lines and cables, in which the current flow out of each conductor flows back by the two other conductors. The inductances and capacitances depend on the geometry of the conductor arrangement or the type of laying of cables in a triangle or in a flat arrangement.

The high-frequency characteristic impedance (wave impedance) differs from this. At high frequency, the return current flows back in cables via the cable sheath and in overhead lines via the neighboring conductors and also via the earth wire and the ground according to the geometry of this conductor arrangement.

In the case of low line losses ($R' \cong 0$, $G' \cong 0$, $\alpha \cong 0$) the phase angle of propagation is:

$$\beta = \omega\sqrt{L'C'} = \omega\sqrt{\mu_0\varepsilon_0\varepsilon_r} \tag{6.5}$$

For overhead lines, the phase angle ($\varepsilon_r = 1$) is:

$$\beta = 2\pi 50\sqrt{1256 \cdot \frac{10^{-6}H}{m} \cdot 8.854 \cdot \frac{10^{-12}F}{m}} = 0.00105 \text{ rad/km} = 0.06°/\text{km}$$

that is 6° per 100 km.

In cross linked polyethylene cables (XLPE cable) due to the permittivity of $\varepsilon_r = 2.3$ the phase shift is 9° per 100 km. This also means that the speed of propagation in XLPE cables is only 66% of that of overhead lines.

Table 6.2 shows the operational characteristic impedance of overhead lines and cables in comparison.

Table 6.2 Operational characteristic impedance of overhead lines and cables

	Voltage level	Conductor	Operational characteristic impedance Z_W
			Ω
Overhead line	110 kV	Single conductor	375
	220 kV	Single conductor	375
	220 kV	Double conductor	240
	380 kV	Double conductor	260
	380 kV	Triple conductor	230
Cable	XLPE cable	Single conductor	20–50

6.3   Operating Behavior of Overhead Lines

Operation with natural power
If a line is terminated with its characteristic impedance Z_W, this represents an operating state with natural power. This is a special case of operation in which the transmitted energy can be fully absorbed at the end of the line and not a part of it is reflected. With natural power, the energy can be transmitted over long distances.

At the end of the line is

$$U_2 = I_2 \cdot Z_W \tag{6.6}$$

This simplifies Eqs. (6.1) and (6.2)

$$U_2 = U_1 e^{-\gamma l} = U_1 e^{-\alpha l} e^{-j\beta l} \tag{6.7}$$

$$I_2 = I_1 e^{-\gamma l} = I_1 e^{-\alpha l} e^{-j\beta l} \tag{6.8}$$

This gives the relationship for the natural power at the end of the line at a distance l from the start of the line

$$P_{nat}(l) = P_2(l) = \sqrt{3} U_2 I_2 = P_1 e^{-2\alpha l} e^{-2j\beta l} \tag{6.9}$$

The efficiency η of the power transmission is

$$\eta = \frac{P_2}{P_1} = e^{-2\alpha l} \tag{6.10}$$

At high nominal voltages, bundled conductors with two and more conductors are used to reduce corona disturbances (pre-discharges on the conductors) (Table 6.3).

For long transmission distances with high transmission powers, high voltages are required. In the European transmission grid, at a voltage of 380 kV, the transmittable natural power per three-phase system is limited to around 650 MVA. This applies, for example, to the transmission of wind energy from northern Germany to southern Germany.

Table 6.3 Natural power of overhead lines

U_n [kV]	Number of conductors n	Z_W Ω	P_{nat} MVA
110	1	375	32
220	1	375	130
	2	240	200
380	3	260	550
	4	230	630
500	3	260	960
750	4	250	2250
1200	8	230	6250

Reactive power balance of the lossless line

With the lossless line ($\alpha = 0$), the line equations are simplified according to Eqs. (6.1) and (6.2).

$$U_1 = U_2 \cos \beta l + jI_2 Z_W \sin \beta l \tag{6.11}$$

$$I_1 = \frac{U_2}{Z_W} j \sin l + I_2 \cos l \tag{6.12}$$

The input impedance of the lossless line is the quotient of these equations.

$$Z_1 = \frac{U_1}{I_1} = \frac{I_2(Z_2 \cos \beta l + j Z_W \sin \beta l)}{I_2(\cos \beta l + j Z_2 / Z_W \sin \beta l)} \tag{6.13}$$

$$Z_1 = Z_W \frac{\frac{Z_2}{Z_W} + j \tan \beta l}{1 + j \frac{Z_2}{Z_W} \tan \beta l} \tag{6.14}$$

From this, the input impedance Z_{1L} of a line for open end (Z_2 infinite) can be determined. In power engineering, only the open circuit is important, but not the short circuit at the end.

$$Z_{1L} = Z_W \frac{1}{j \tan \beta l} = -j Z_W \cdot \cot \beta l \tag{6.15}$$

In the case of overhead lines, the input impedance of the lossless and no-load line is capacitive in the range of length from zero to 1500 km. At 1500 km the line is in resonance (a quarter of the wavelength of 50 Hz). The input impedance of the lossless overhead line is then zero, since with this line length the open circuit at the end is transformed into a short circuit at the input. In the range from 1500 to 3000 km, the input impedance is inductive. At 3000 km—this is the half wavelength—the input impedance is infinite, as the output impedance is transformed unchanged on the input impedance.

When operating with natural power ($Z_2 = Z_W$), according to Eq. (6.14)

$$Z_1 = Z_W \tag{6.16}$$

The input impedance is then constant and no longer dependent on the line length.

Operation at the thermal power limit

For short transmission distances overhead lines can be operated at the thermal power limit. This is determined by the current carrying capacity of the conductors, which according to the EN 50182 standard apply to an ambient temperature of 35 °C and a wind speed of 0.6 m/s [Fischer 1989], [Niemeyer 1992].

Table 6.4 shows the thermal limits of different overhead line variants [Brauner 2017]. Three variants are shown for the 220 kV overhead lines. The transmission line "220 kV old" represents a formerly common design with single conductors and the conductor cross-section of 240 mm² made of aluminum for the current carrying

Table 6.4 Current carrying capacity and thermal power limit of overhead line conductors according to EN 50182

Nominal voltage	Conductor type	Current carrying capacity According to EN 50182 35 °C, 0.6 m/s	Thermal power 35 °C, 0.6 m/s
110 kV	243-AL1/39-ST1A	645 A	137 MVA
220 kV old	243-AL1/39-ST1A	645 A	273 MVA
220 kV HQ	341-AL1/109-ST1A	800 A	339 MVA
220 kV ZTAL	*341/109*	*(210 °C) 1566 A*	*665 MVA*
380 kV	3 × 679-AL1/86-ST1A	3627 A	2638 MVA

capacity of 645 A and a steel core of 40 mm² for the mechanical bearing capacity. In order to increase the transmission capacity, the conductors can be exchanged.

In the "220 kV HQ" variant, the cross-section is increased to 341 mm² aluminum and 109 mm² steel core, thus increasing the current-carrying capacity to 800 A. Alternatively, the conductors can be exchanged for high-temperature conductors in the *"220 kV ZTAL"* variant. These allow an increased conductor temperature of 210 °C, instead of the design according to the standard with 80 °C. However, the increased current-carrying capacity of 1566 A is associated with higher line losses by the factor of 2.5. With high-temperature conductors, the line capacity can almost be doubled from 800 to 1566 A.

In the long term, significantly larger grid capacities must be created to connect wind energy. A conversion of the overhead lines from 220 to 380 kV results in an almost identical appearance of the overhead lines with roughly the same mast height and the same width of right of way and creates significantly larger transmission capacities of 2638 MVA here with low network losses. In order to minimize corona disturbances at the conductors, 380 kV overhead lines are almost exclusively designed with conductor bundles as a triple or quadruple bundle [Oeding 2004].

Table 6.4 shows a layout of a 380 kV overhead line with three bundles 3 × 679/86. This variant outperforms all other variants with its current carrying capacity of 3627 A at a thermal limit power of 2638 MVA and at the same time normal conductor temperatures and low transmission losses. This is the technical and economical best variant. Its thermal transmission capacity is equivalent to eight parallel 220-kV transmission lines per 339 MVA, or to the capacity of four parallel 220-kV transmission lines with high-temperature conductors.

Since large capacities can thus be created on an existing 220-kV routes and no parallel lines are required, the conversion of the transmission systems in their 220 kV remnants to 380 kV represents a very efficient and resource-saving measure. This allows significant capacities to be created for the integration of large wind farms on the existing line routes.

Currently, the unit power of wind turbines is in the range of 3 to 5 MW and transmission capacities of several 1000 MVA are required to connect large wind farms. It is immediately apparent, that only by switching to 380 kV with bundle conductors, energy transport in the order of the medium-term expansion of wind parks of several thousand MVA transmitted, will be possible.

Since 220 kV overhead lines account for around 60 to 80% of the investment and operating costs of 380 kV overhead lines, parallel 220 kV overhead lines or conversion of these to high-temperature lines are significantly less economical in the long term. In addition, parallel 220 kV overhead lines on new routes are unlike to be approvable compared to significant increases in capacity on existing routes due to the conversion to 380 kV, since 220 kV represent an inefficient use of space. Finally, even if the entire transmission system is converted to 380 kV, the existing main transformer from 380 to 220 kV are no more needed, which, with their rated outputs of 600 MVA each, also represent a transmission bottleneck. They are also a safety risk in the event of failure because of their long repair times.

6.4   Operating Behavior of Cables

6.4.1   Technical Properties of Cables

Cables are available for all voltage levels. They allow underground installation. In the distribution networks of low and medium voltage, they can be used cost-effectively and their technical properties are well suited for these supply tasks. In the high-voltage and extra-high-voltage levels, their technical properties are rather limited compared to overhead lines and they are only used there in special cases, especially for underground supply of cities over short distance.

The inner conductors are often made of aluminum for voltages up to 110 kV. At 380 kV, copper is mainly used because the power loss and the associated heat dissipation represent a technical problem here due to the thick-layer insulation.

As the insulating medium in the low voltage range Polyvinyl chloride (PVC) and cross-linked Polyethylene (XLPE) are used. Because of the low dielectric losses, only XLPE is used in the medium-voltage level and for all higher voltages. Oil-paper insulation has historical significance and is now found only in special cases, especially at high DC voltages. Polyethylene has a permittivity of $\varepsilon_r = 2.3$, resulting in lower cable capacitances compared to oil-impregnated paper ($\varepsilon_r = 3.5$) and PVC insulation ($\varepsilon_r = 4.0$) [Kabel-HB 1986], [Kiwit 1985].

6.4.2   Thermal Operating Behavior of Cables

In the inner conductor of cables, the current produces heat loss. In normal operation, these are caused by the load currents, or in faulted operation by the short circuit current. The temperature of the inner conductor should not exceed 90 °C in normal operation with XLPE, as this shortens the service life of XLPE cables.

Short-circuit processes can produce high currents for a short time. Only short-term temperatures of 250 °C are permitted with XPLE. PVC cables only allow operating temperatures of 70 °C and short-circuit temperatures of 140 to 180 °C.

Another cause of the thermal stress on cables are the dielectric losses of the insulating material. They increase with the square of the voltage. XLPE has very low losses (tan $\delta = 0.0004$) compared to PVC (tan $\delta = 0.05$) and oil-paper insulation (tan $\delta = 0.003$). The high losses of PVC only allow use in low voltage.

The cable losses of the inner conductor and the insulation must be dissipated via the insulation. This poses a problem as materials with good electrical insulating properties generally also have poor thermal conductivity.

At higher operating voltages, heat dissipation increasingly determines the structural design of cables. In the case of extra-high voltage cables, copper with larger cross-sections is used for the inner conductor to reduce heat loss. Furthermore, in order to improve the heat dissipation via the insulation, the insulation thickness of these cables is less than proportional in relation to the higher voltage compared to medium-voltage cables. Due to the relatively lower insulation thickness, the field strength in the insulating body is higher, which means a higher dielectric stress and thus a shorter service life.

Figure 6.1 shows the typical structure of XLPE cables. Around the inner conductor (here made of aluminum), a thin layer of conductive Polyethylene is first applied, to improve the field strength on the surface of the inner conductor. Above it lies the light-colored insulating body, the thickness of which is designed according to the nominal voltage. On top of this, is again a layer of conductive polyethylene. Ribbons are wrapped over this and the cable shield made of copper wires and a copper tape follows. Finally, a foil is applied and the hard polyethylene jacket follows.

The cables are laid in the ground. The cables can be arranged close together in a triangle. This results in poorer heat dissipation into the ground. Often, cables are laid side by side in one plane with sufficient spacing, which results in better heat dissipation.

The losses P'_V in the inner conductor per unit length of the cable are caused by the resistance of the inner conductor R' and the current I.

$$P'_V = R' \cdot I^2 = \frac{I^2}{\kappa A} \tag{6.17}$$

Fig. 6.1 Structure of a XLPE cable for medium voltage (ÖKW)

By designing the inner conductor with a large cross-section A and using material with a high specific electrical conductivity κ, low losses can be achieved. In medium voltage power cables because of the low thickness of the XLPE insulation layers aluminum [$\kappa = 35.4$ m/(Ω mm^2)] can be used as the inner conductor. Copper is commonly used for extra-high voltage cables [$\kappa = 56$ m/(Ω mm^2)].

The dielectric losses P'_d per phase and unit length result from

$$P'_d = \frac{U_n^2}{3}\omega C' \cdot \tan \delta \qquad (6.18)$$

where C' represents the capacitance per phase.

The losses in the inner conductor must penetrate the entire insulating space, while the dielectric losses occur in the insulating body and therefore only have to penetrate half the space on average.

The increase in the conductor temperature ΔT is determined by the sum of thermal resistances R'_W from the cable insulation and the surrounding soil

$$\Delta T = R'_W \cdot \left(P'_V + P'_d\right) \leq T_{max} \qquad (6.19)$$

and should in continuous operation be significantly lower than the temperature limit of 90 °C of XPE. If the individual phases are laid in horizontal arrangement at a sufficient distance from each other and at sufficient depth, the individual thermal resistances can be understood as coaxially layered cylinders, consisting of the concentric cylinders of insulating material, of the cable shield and the surrounding soil. Since the heat transport of the cable causes the soil to dry out in the vicinity of the cable, the soil should be simulated by several cylindrical bodies with different soil moisture and thus different thermal conductivity.

The specific thermal resistance is related to a cable length of one meter and, in the case of cylindrical arrangements, is to be calculated as follows for each partial cylinder i

$$R'_{wi} = \frac{\rho_{wi}}{2\pi}\ln\frac{r_{2i}}{r_{1i}} \qquad (6.20)$$

where ρ_{Wi} is the specific thermal resistance of a cylinder with the dimension K·m/W and r_{1i} represents the inner radius of the cylinder i and r_{2i} the outer radius. XLPE has with $\rho_{W,XLPE} = 3.5$ K·m/W a lower heat resistance than PVC with $\rho_{W,PVC} = 6.0$ K·m/W.

Table 6.5 Specific thermal resistance of the soil	State of the soil	Specific thermal resistance
	Wet	$\rho_W = 0.7$ K·m/W
	Normal wet	$\rho_W = 1.0$ K·m/W
	Planning value for cables	$\rho_W = 1.5$ K·m/W
	Dry	$\rho_W = 2.0$ K·m/W
	Dried out	$\rho_W = 3.0$ K·m/W
	Extremely dried out	$\rho_W = 5.0$ K·m/W

Fig. 6.2 Influence of soil conditions on the load capacity of cables

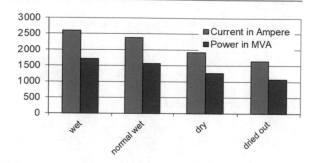

The thermal resistance of the soil depends on the moisture content (Table 6.5).

Figure 6.2 shows the load capacity of a 380 kV XLPE cable with a copper conductor of 2500 mm² with different soil moisture contents.

Longer periods without sufficient precipitation can therefore reduce the transmission capacity of cables. In New Zealand [Auckland 1998] in a prolonged dry period at high temperatures, the soil was extremely dry. At the same time, the demand for electricity was high due to the increased use of air conditioning. The decreasing soil moisture led to overheating of the four 110 kV cable systems with oil-paper insulation. One after the other, all four cable systems failed and the major city of Auckland was without electrical supply. Since it is not possible to produce and lay $4 \times 3 \times 10$ km $= 120$ km of cables at short notice, the power supply was interrupted for eight weeks. The restoration was initially carried out via an emergency overhead line along the railway line.

Example: heat calculation for a 380 kV cable

A XLPE cable for 380 kV with a copper conductor cross-section of 1200 mm² and an outside diameter of 90 mm is to be used to supply a large city. The dimensions are for the inner conductor radius $r_1 = 19.5$ mm, inner radius of the outer conductor $r_2 = 40$ mm and outer radius $r_3 = 45$ mm. The surrounding soil is assumed to be dry $\rho_{WS} = 2.0$ Km/W. The thermal resistance of XLPE is $\rho_{W,XLPE} = 3.5$ Km/W. The copper shield of the cable is not taken into account because of its good thermal conductivity.

Question: What current load on the inner conductor must not be exceeded in continuous operation so that its temperature does not exceed 80 °C?

a) Calculation of the specific electrical resistance of the inner conductor:

DC resistance:

$$R'_{=} = \frac{1}{\kappa A} = \frac{1}{35\frac{m}{\Omega \cdot mm^2} 1200 \ mm^2} = 2.4 \cdot 10^{-5} \Omega/m$$

The AC resistance is larger because of the skin effect and is about $3.0 \cdot 10^{-5}$ Ω/m. Converted to the conductor limit temperature of 80 °C, the alternating current resistance is $3.7 \cdot 10^{-5}$ Ω/m.

Calculation of the specific thermal resistances
Insolation of the cable:

$$R'_{w1} = \frac{\rho_{W,XLPE}}{2\pi} \ln \frac{r_2}{r_1} = \frac{3.5}{2\pi} \ln \frac{40}{19.5} = 0.4 \left[\frac{K \cdot m}{W} \right]$$

Outer sheath of the cable:

$$R'_{w2} = \frac{\rho_{W,XLPE}}{2\pi} \ln \frac{r_3}{r_2} = \frac{3.5}{2\pi} \ln \frac{45}{40} = 0.066 \left[\frac{K \cdot m}{W} \right]$$

Surrounding soil (dry) is calculated up to a radius of 50 cm:

$$R'_{w3} = \frac{\rho_{W,soil}}{2\pi} \ln \frac{r_4}{r_3} = \frac{2.0}{2\pi} \ln \frac{500}{45} = 0.766 \left[\frac{K \cdot m}{W} \right]$$

The soil has a significant impact once it has dried out. When laying cables, one tries to prevent this by using special sand or a lean concrete fill around the cable.

The total thermal resistance is:

$$R'_W = R'_{W1} + R'_{w2} + R'_{w3} = 1.232 \left[\frac{K \cdot m}{W} \right]$$

For the permissible temperature increase of 60 °C from 20 °C to the limit tempera-ture of 80 °C, the permissible power loss can be calculated.

$$P'_v = \frac{\Delta T}{R'_w} = \frac{60°C}{1.232} = 48.7 \frac{W}{m}$$

This power loss can be dissipated with the ground conditions per meter of length of the cable. From the AC resistance also the permissible phase currents and the three-phase AC power can be determined.

$$P'_v = R' \cdot I^2$$

$$I = \sqrt{P'_v / R'} = \sqrt{\frac{48.7\,W/m}{3.7 \cdot 10^{-5}/m}} = 1147\,A$$

This corresponds to a three-phase transmission power of 755 MVA. The heat losses of neighboring cables were not taken into account due to horizontal arrangement at distance. In shorter distance, this can reduce the power by 20 to 30%.

6.4.3   Properties of Medium Voltage and High Voltage Cables

Energy cables have become very popular in the area of low and medium voltage (Table 6.6).

In Austria, the relative portions show that 80.6% of the low-voltage system lengths and 62.2% of the medium-voltage systems are now cabled. In the medium and low voltage level, cables can often be installed cost-effectively at depths of

Table 6.6 System lengths of overhead lines and cables in Austria [e-control 2016]

Voltage level	Overhead lines		Cable		Total	Voltage level
	km	portion	km	portion	**km**	Relative share of cables
380 kV	2 996	1.1%	55	0.0%	**3051**	1.8%
220 kV	3 705	1.4%	8	0.0%	**3713**	0.2%
110 kV	10,662	4.1%	774	0.3%	**11,436**	6.8%
1 kV–<110 kV	26,122	10.0%	42,940	16.5%	**69,062**	62.2%
1 kV and below	33,616	12.9%	139,753	53.6%	**173,369**	80.6%
Total	**77,102**	**29.6%**	**183,529**	**70.4%**	**250,631**	**70.4%**

up to two meters using a laying plow. Since the load capacity decreases with the installation depth, depths of 0.7 to 1.1 m are often chosen.

Also, three cables including warning tape can be laid in a single plowing pass. The daily length is around 3 km. Compared to the classic excavation method, there is no mixing of the soil layers in the excavated material. The method is in particularly suitable for valley areas with natural soil.

In the field of high and extra-high voltage cables, however, elaborate excavation work is necessary with the introduction of special bedding material for thermal stabilization of the cable. This changes the soil structure permanently.

Compared to overhead lines, cables have significantly larger capacitances C'. As a result, reactive power is generated when operating with AC voltage. The reactive power Q increases proportionally to the capacitance C' of the cables and quadratic with the nominal voltage Un.

$$Q = U_n^2 \omega C \tag{6.21}$$

This complicates the use of cables, especially in the high and extra-high voltage range. When operating with natural power, the inductive component of the cable just compensates for the capacitive component and the input resistance of the cable corresponds to the characteristic impedance. Since the characteristic impedance of cables is very low, the natural power of cables is far above their nominal thermal capability. They have a capacitive behavior in normal operation.

$$P_{nat} = \frac{U_n^2}{Z_w} \tag{6.22}$$

380 kV overhead line:

$$P_{nat} = \frac{U_n^2}{Z_w} = \frac{380^2}{243} = 594 \, \text{MVA}$$

380 kV cable:

$$P_{nat} = \frac{U_n^2}{Z_w} = \frac{380^2}{52} = 2777 \, \text{MVA}$$

Table 6.7 Characteristic values of overhead lines and cables for 380 kV

	Overhead line	Cable
Nominal voltage	380 kV	380 kV
Cross-section	3 × 679-AL1/86-ST1A	1200 Al
Capacity	0.014 μF/km	0.180 μF/km
Thermal current limit	3627 A	1245 A
Operational charact. Impedance	243 Ω	52 Ω
Natural power	594 MVA	2777 MVA
Thermal power limit	2410 MVA	825 MVA

In the case of overhead lines, on the other hand, self-compensation is achieved in the normal operating range, which is below the thermal limit power. In Table 6.7 are the data of a 380 kV overhead line and a 380 kV cable comparatively shown.

As shown in formula (6.14), the input impedance of an overhead line or cable, operated at natural power $(Z_2 = Z_W)$, corresponds to the characteristic impedance and no reactive power is consumed. Since cables must always be operated well below their natural power, due to their low characteristic impedance, they show a capacitive behavior over the entire operating range. This limits their useable length in the range of high and extra-high voltage, since the generated reactive power according to Eq. (6.21) increases quadratically with the operating voltage. Extra-high voltage cables must therefore be compensated at regular intervals by connecting inductances in parallel.

Figure 6.3 shows the reactive power demand for the overhead line (top) and the cable (bottom) of Table 6.7 [Oswald 2007 and 2010]. The lengths of overhead line and cable are 30 km here.

The overhead line compensates itself at 594 MVA. This value is below the thermal limit power of 2410 MVA. The 380 kV cable has a thermal power limit of 825 MVA and can therefore never transmit the natural power of 2777 MVA.

The high reactive power consumption of cables is also the reason for their limited transmission capability. The transmittable current I_W is reduced by the reactive power consumption I_b compared to the nominal current I_n.

Fig. 6.3 Reactive power requirement of an overhead line and a cable for 380 kV

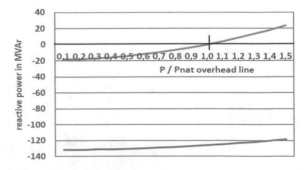

$$I_W = \sqrt{I_n^2 - I_b^2} \tag{6.23}$$

Figure 6.4 shows this for cables of medium, high and extra-high voltage. Oil-paper cables for 380 kV have a higher permittivity of the insulating material than XLPE and therefore have higher cable capacitances and thus shorter transmission lengths. Since the reactive current consumption decreases quadratically with the operating voltage, medium-voltage cables for 10 to 30 kV can also be used without compensation in the range of the expansion of medium-voltage networks, which extend up to about 50 km.

Cables for 380 kV and 110 kV are found almost exclusively in urban areas. They are then used for short transmission distances on the order of 5 to 10 km. Many cities have two to four high-voltage or extra-high voltage feed-in stations from which the high-voltage cables lead into the city center. The cables are not connected through between the feeder stations. This ensures that there is no over-load due to transmission via the cables, but that they can only be loaded with the load current of the urban substations.

Because of the long repair times of extra-high voltage cables compared to over-head lines, at least two cable systems are generally used per overhead line system. At 380 kV, the costs of cable routes are 5 to 10 times higher than those of overhead lines.

Medium voltage cables can be laid in longer lengths without intermediate joints. Due to the lower weight, transport lengths of up to 20 km are possible with a total transport weight of 40 t. Figure 6.5 shows the possible transport lengths for the individual voltage levels. The long transport lengths enable efficient laying using a cable plow.

In the case of XLPE cables for 380 kV, the transport lengths are less than one kilometer, which means that a large number of joints have to be used to connect these short cable sections. In the case of XLPE cables, prefabricated joints are pre-dominantly used, which are to be installed on site under clean room conditions. Faults in cable joints are the most common internal cause of defects. The most

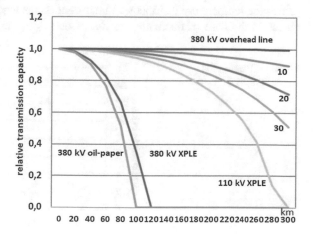

Fig. 6.4 Comparison of transmission capability of overhead lines and XLPE cables

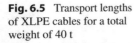

Fig. 6.5 Transport lengths
of XLPE cables for a total
weight of 40 t

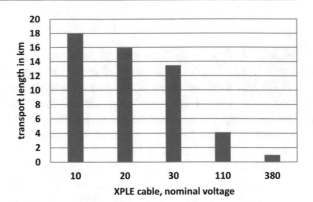

common external causes of defects result in excavation work in which the cable
jacket is damaged.

The heat dissipation in cables limits their transmission capacity. In order to
maintain sufficient transmission capacity in the range of high nominal voltages,
the insulating layers of cables are not enlarged proportionally to the nominal voltage. In the case of extra-high voltage cables, this means that the specific operating field strength is higher than in the area of medium-voltage cables. Figure 6.6
shows the operating field strength of XLPE cables.

Cables for 380 kV have about three times the electric field strength of medium
voltage cables. High field strengths cause the insulation material to age faster and
larger insulation volumes increase the risk of defects. According to the current
state of technology, the service life of XLPE cables is 40 to 50 years.

6.4.4   Dynamic Operating Behavior of Cables

So far, the heating of cables has been investigated using a static model. This determines the temperature distribution under continuous load. To determine the temperatures in the case of fluctuating loads, as occurs in the transmission of renewably

Fig. 6.6 Electric field
strength in XLPE cables

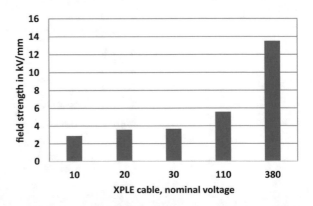

generated electricity, a dynamic model is necessary. This involves modeling the thermal resistances within the cable and in the surrounding soil and the heat capacities of the individual layers. As long as the individual cables are laid at a sufficient distance, coaxial cylindrical layers can be modeled. With increasing distance from the axis of the cable the thermal resistance decreases and the heat capacity increases according to the increase in surface area and volume of the equidistant cylindric layers.

Small overloads of the cable result in small gradients for the temperature rise, due to the large thermal capacity of the surrounding soil.

In central Europe at the laying depth of cables, which is 0.7 to 2 m, the temperature range is 5 to 15 °C, when uninfluenced by cable heat [Brakelmann 1985], [Siemens 1987]. The lowest temperature values occur in the period from January to April and the higher ones in the range from June to October (Fig. 6.7).

In countries with a high mean annual temperature such as Kuwait, the soil heats up significantly more in the laying depth of cables, which means that the mean cable load must be reduced in the planning.

The thermal resistance of the soil depends on its ability to bind water. Sandy soils show poor thermal conductivity, while clay soils are favorable. Fig. 6.8 shows the annual variation of the thermal resistance of these soils [Siemens 1987].

Figure 6.9 shows the heating of a 380 kV cable with a copper conductor of 2500 mm^2 at normal soil conductivity of 1.0 K·m/W. The surrounding soil results

Fig. 6.7 Soil temperatures in Germany and Kuwait

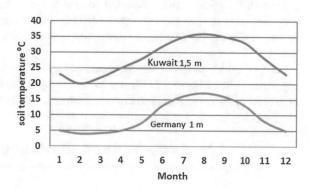

Fig. 6.8 Annual variation of the thermal resistance of clay and sandy soil

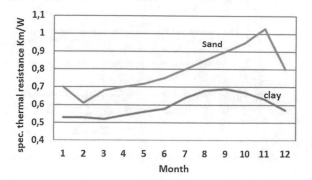

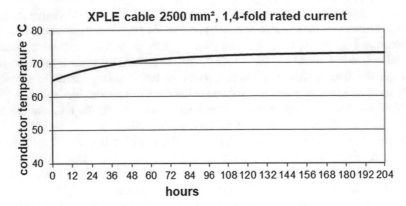

Fig. 6.9   Heating up of a 380 kV XLPE cable due to permanent overload

in large heat capacities and a large time constant for the heating of the entire system of cables and soil.

At the beginning, a soil temperature of 20 °C was assumed in the vicinity of the cable. If the load has a longer duration, a constant thermal gradient is established in the surrounding soil.

The large time constant of the heating turns out to be a disadvantage when the cable has to cool down after a long period of high load. Since the surrounding soil is heated, it makes it more difficult for the cable to cool down when the current load has decreased to the rated current. This is illustrated in Fig. 6.10. In the event of major disturbances in the transmission system, high line loads can temporarily occur.

In the 380 kV transmission network, cables are used almost exclusively to supply large cities. Figure 6.11 shows as an example the transmission network for a large city with 2.5 million inhabitants. Power is fed in via a northern and southern

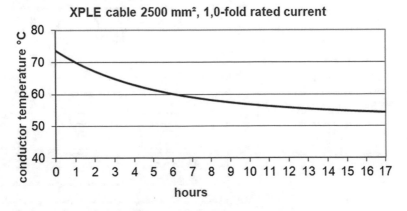

Fig. 6.10   Cooling of a 380 kV XLPE cable at rated current after overload

Fig. 6.11 380 kV cable
transmission system in a
large city with 2.5 million
inhabitants

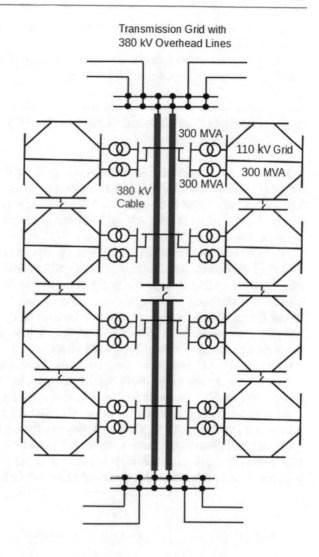

connection to the superimposed 380 kV transmission grid. In urban areas, energy
is transmitted with 380 kV XLPE cables. These are separated in the middle. This
means that transmission and overloading in the event of major external disturbances are not possible; only the load currents of the urban supply area can occur
as a load. If the northern or southern grid connection fails, the 380 kV cables
can be interconnected and full supply can be temporarily provided from one of
the network connections. The urban area is fed via underlying 110 kV cable networks with a diameter of around one kilometer. These are designed for outputs of
300 MVA each and are fed by two transformers of 300 MVA each. The failure of
a transformer does not lead to a supply interruption. In case of failure of the doubled transformers, supply from neighboring 110 kV networks is possible by means

of switching operations. Such network structures are characterized by a very high supply reliability.

The diameter of the metropolitan network is around 10 km, so the unconnected 380 kV cable sections are around 5 km long.

6.5   Connection of Wind Farms in the High and Extra-High Voltage Grid

Wind farms are integrated into transmission grids with 380 kV and in high voltage distribution networks with 110 kV. While the power of the wind farms in the 110 kV grid is limited to around 200 MVA per network group, several 1000 MW can be integrated in the 380 kV grid via correspondingly powerful line connections per wind farm.

The wind turbines themselves are initially connected over a small area using 30 kV XLPE cables. An aluminum cross-section of 300 mm^2 results in a transmission capacity of 25 MVA, and with 500 mm^2 results in 32 MVA. This means that 8 to 10 wind turbines, of 3 MVA each, can be connected to the grid per cable run. In small-scale connected wind farms, the generation characteristics of the individual plants are very similar as a result of the wind supply. Via 110 kV grids, the medium-voltage cable networks of wind farms can be combined to produce larger power. When wind energy is connected, the local 110 kV grids require higher transmission capacity to the overlying 380 kV grid, to enable to transport the wind energy. As Fig. 6.12 shows, the expansion of wind energy in eastern Austria has led to peak generation capacities that can be up to four times the peak load of the local power grid [Sinowatz 2015]. Without high-capacity substations to the transmission grid, wind turbines would have to be curtailed. Figure 6.15 shows the connection of the wind farms in eastern Austria via the newly built Styria line. Hereby the surplus power can be distributed across Austria and thus be used [Sinowatz 2015].

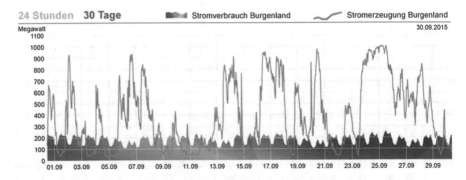

Fig. 6.12   Wind generation and electricity demand in Burgenland [Sinowatz 2015]

In the following, a comparison of the integration of high-power wind farms via cables or overhead lines for 110 kV and 380 kV will be carried out.

The initial design of 110 kV cables for the connection of wind farms according to utility load with a load factor of 0.7 has proven to be inappropriate and led to frequent cable damage due to overloading.

When designing for utility load, it is assumed that the peak load of the day is followed by the low load of the night and that the cables can cool down during this period. Operational experience with wind energy has shown that several days with full generation capacity from wind energy plants are possible. The cables for the transport of wind energy must therefore be designed for continuous load with a load factor of 1.0.

The load capacity of cables has already been investigated. It depends on the thermal conductivity of the soil and thus on the soil moisture. With overhead lines, the transmission capacity depends on the ambient temperature and the wind speed. The nominal transmission capacity is valid for an ambient temperature of 35 °C and a wind speed of 0.6 m/s and for a maximum conductor temperature of 80 °C. The wind speed of 0.6 m/s corresponds to the rising air caused by soil thermals.

Wind turbines are designed to deliver full power at a wind speed of 10 to 12 m/s. The highest wind production takes place in winter with low ambient temperatures and the mean annual ambient temperature is 15 °C in Austria.

The model of the thermal balance of the conductors of overhead lines consists of the following four components

- Heat loss in the conductor. This grows in proportion to the square of the operating current.
- Heating of the conductors by solar radiation. This depends on the projected absorption area to the sun and the degree of absorption of the conductor surface.
- Heat dissipation through thermal radiation. This is determined by the total surface area of the conductors and the radiation properties of the conductor surface.
- Heat dissipation through convection. This is determined by the relative wind speed at the conductor surface.

The conductor temperature is determined by the thermal equilibrium of heating and heat dissipation and must not exceed 80 °C in normal operation. Using a thermodynamic model, which takes into account all the influencing parameters shown above, the overload capacity of overhead conductors was calculated at different ambient temperatures and wind speeds. The solar radiation is 900 W/m^2. The conductor has the cross-sections 679-AL1/117-ST1A and is designed for a rated current of 1150 A at a conductor temperature of 80 °C. This current applies to the standardized design in accordance with EN 50182 at an ambient temperature of 35 °C and a wind speed of 0.6 m/s.

This conductor is used in some new Austrian 380 kV overhead lines in the form of a triple bundle. Figure 6.13 shows the overload capacity of this conductor at different ambient temperatures and wind speeds. The red point in Fig. 6.12 represents the nominal load capacity according to the standard.

With the nominal power of the wind energy plants at 12 m/s in winter, operation at an ambient temperature of 10 °C, overhead lines with this type of conductor achieve three times the nominal power. At 20 °C, it is still a factor 2.7 and at the extreme temperature of 35 °C, the factor 2.3. The most common wind speed in areas with wind generation is 4 to 6 m/s. Here, the wind turbines only achieve less than 15% of their nominal output, due to the cubic dependence on the wind speed. The capacity of the overhead lines corresponds here to twice the rated power.

The high overload capacity of overhead lines at higher wind speeds represents a major advantage for the connection of wind energy. With buried cables, the thermal limiting current remains unchanged at higher wind speeds.

Under overload conditions, overhead lines have smaller time constants for heating and cooling than cables. Figure 6.14 shows a heating process under the nominal conditions of 35 °C ambient temperature and 0.6 m/s wind speed. Here, starting from a load that results in a conductor temperature of 60 °C, the conductor

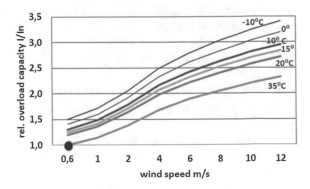

Fig. 6.13 Overload capacity of a 679-AL1/117-ST1A conductor at the maximum permissible conductor temperature of 80 °C

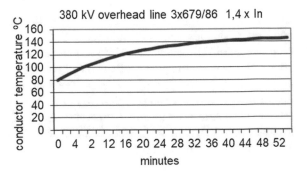

Fig. 6.14 Heating up an overhead line with 1.4 times the nominal current

is suddenly loaded with 1.4 times the rated current. The time constant for heating is 20 min, as shown in Fig. 6.14.

The same time constant also applies to cooling. An overhead line is therefore available for grid operation again more quickly after an overload.

According to Fig. 6.13, an overload with a factor of 1.4 and a wind speed of 2 m/s would again reach a conductor temperature of 80 °C even at 35 °C. At lower ambient temperatures, the conductor limit temperature would even be reached already at around 1 m/s.

The higher capacities of overhead lines can only be used if the overcurrent protection assigned to the lines can be adjusted according to the weather conditions based on wind speed and ambient temperature. Recent studies by the transmission system operators are moving in this direction and attempting to set the overcurrent protection more flexibly by means of remote parameterization. The so-called *"Dynamic Thermal Rating of Overhead Lines"* [Jupe 2011] can only be used to a limited extent. It assumes that the existing switchgear is also capable of correspondingly overloading. Further, in the transmission system also high loads are possible in the event of faults or remote wind production, while at the same time the local ambient temperature can be high and the wind speed low. Under these conditions, line overloading may occur and congestion management cannot be dispensed with. A prerequisite for optimal use of overhead lines is an area-wide monitoring of ambient conditions and line loads.

From an operational point of view, replacing old overhead lines by new ones with sufficient capacities is the more suitable measure in the long term, since the statically available higher line capacities then permit simpler and more reliable operation without frequent grid congestion situations.

When connecting wind farms to the transmission network, the proactive behavior of the overhead lines at the wind farm can be used. In particular if one line system fails when there is high wind, the remaining system can be operated with twice the capacity. For the connection of wind farms in the area of high and extra-high voltage, overhead lines are therefore the economically and technically best option.

6.6   Safety and Reliability of Overhead Lines and Cables

The mean failure rates and repair times define the reliability of overhead lines and cables. For new installations, the failure frequency is initially determined by manufacturing defects and errors in the installation. During the service life, these are internal faults due to overload and aging phenomena or external faults by atmospheric overvoltages, natural disasters or human error, that lead to failures. At the end of service life, it is aging, leading even at low overvoltage to failures. External faults due to excavation work are a frequent cause of faults throughout the entire service life of cables.

Overhead lines use as insulators porcelain, toughened glass or glass-fiber-reinforced polymer insulators on silicon rubber basis. These insulators are equipped with protective rings in order to achieve a flashover of the air gap in the event of over-voltages and to avoid a surface flashover or a breakdown of the solid insulator [Hütte 1988]. The air insulation of the overhead lines is self-healing. After a lightning strike, which leads to a breakdown of the air insulation, the resulting arc is extinguished by means of an automatic reclosure. Here, the affected phase is disconnected on both sides for about 0.7 s and then reconnected.

In case of cables, the insulating body is permanently damaged in the event of a breakdown. After a fault has been located, the affected cable section must be cut out and replaced with a similar one using two joints. Another cause of failure is excavation work that leads to damage. This can cause water to penetrate and, in the case of XLPE cables, lead to a progressive degradation of the insulation through so-called "water treeing".

For overhead lines, due to the large inventory of systems lengths and the long observation period since the introduction of the 380 kV voltage level in 1950, the failure and repair rates are available from fault and damage statistics.

380 kV cables are mainly found only in urban transmission networks, where they have typical system lengths of 5 to 10 km. The previously used oil-paper cables are being displaced by XLPE cables in new installations. The statistical characteristic values of the cables still vary widely between the individual projects and there are no reliable values based on long-term observation of large collectives. The mean unavailability is defined in the reliability calculation by the failure rate λ and the mean repair rate μ per 100 km system length and per year. The unavailability NV in hours per year is defined by the failure rate and the mean repair time T_R.

$$NV = \frac{\lambda}{\mu + \lambda} \approx \lambda \cdot T_R \qquad (6.24)$$

For XLPE cables, a failure rate of 0.7/100 km/a was estimated for future systems and with a successful learning curve during assembly (Table 6.8). XLPE cables for 380 kV have very long repair times of 2 to 3 months per damage. Several parallel XLPE cable systems are required to achieve a comparable low unavailability as for overhead lines.

Table 6.8 Failure rates and unavailability of cables and overhead lines at 380 kV

380 kV	Failure rate λ	Mean repair time T_R	Unavailability	Unavailability
	per 100 km·a		per 100 km·a	h/a
Overhead line	0.353	3 h	0.0003356	1, 04
Oil-paper cables	0.595	2–3 months	0.124	1085
XLPE cables	0.7	2–3 months	0.146	1278
2 × XLPE cables	0.102	2–3 months	0.0213	186
3 × XLPE cables	0.0149	2–3 months	0.00311	27

6.7   Energy Economy of Grids

6.7.1   Approval Procedure and Environmental Impact Assessment

The new construction or replacement of overhead lines and cables is subject to an extensive environmental impact assessment. The following must be checked for the project:

- Energy economic necessity
  Since lines have a long service life, the need for energy-economics necessity must be justified, in particular with regard to long-term anticipated developments in electrical energy supply. Long-term strategies of the EU and national energy strategies are to be taken into account. The necessity is to be justified from the future types of use such as integration of renewable energies into the grid, local or national electricity transport and international energy exchange with neighboring countries as well as the development of the grid load.
- State of the art
  The proposed project should correspond to the state of the art from technology from an energy perspective, i.e., efficient in terms of its investment requirements and have sufficient service life, operational hours of use, and contribute to security of supply.
- Alternative solutions
  Alternative solutions must be examined, such as upgrading existing overhead lines by installing thicker conductor cross-sections or high-temperature conductors. Underground cables or future decentralized energy supply concepts can also be considered here. All alternative solutions are to be assessed comparatively according to technical and energy-economic criteria. In particular, investment and operating costs, compatibility with existing technical systems, security of supply and environmental compatibility are important.
- Zero variant
  In this context, it is examined whether a new construction can be dispensed with, due to the state of preservation and the possibilities of upgrading existing systems, or which technical and energy-economic consequences would arise if a new construction were dispensed with.
- Public interest
  If the long-term energy-economic necessity is given, the current and expected criteria of the national and European energy strategy are fulfilled and no suitable alternative solutions or zero variants are given, there is a public interest for a project.

The economic review of the project is not part of the approval procedure. The operator is responsible for assessing the economic viability of a project.

6.7.2   Energy Economic Analysis of Security of Supply

Energy grids are subject to constant adaptation and expansion in line with the new tasks of integrating renewable energy, phasing out of nuclear power and limitation coal-fired generation, or replacing power plant capacity and the increased demand from new end-use applications such as heat pumps and electric vehicles. This results in a different use and load on the existing grid structures.

The expansion of the networks should take place as evenly and uniformly as possible within a network system. Operation is simplified if all lines have similar electrical properties and overload capacities and there are no bottlenecks due to areas with low transmission capacities or different operating behavior.

If, for example, the surrounding grids are upgraded to higher voltage levels and higher transmission capacities and existing networks cannot be expanded in a corresponding timely manner, overloads of this grids are possible due to disturbances in the surrounding grids. Figure 6.15 shows the expansion status of the Austrian network in 2015 (APG, BEWAG).

There are many 220 kV lines (green) in the north–south direction. The 380 kV overhead line (red) in the east from Vienna to Kainachtal could be realized after 22-year approval procedure. It is necessary for the connection of wind energy in Burgenland (yellow area) and Lower Austria. It also improves the fault tolerance in case of strong north–south transmission power flow in the European transmission grid.

The planned conversion of the overhead lines from 220 to 380 kV from St. Peter to Tauern could only be implemented as far as Salzach, due to the long-term

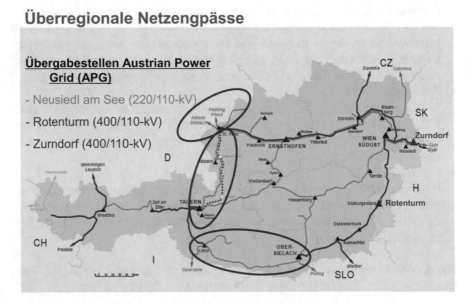

Fig. 6.15   State of expansion of the Austrian network in 2015 [Sinowatz 2015]

approval process. It is necessary to connect the pumped storage power plants in Tauern in order to be able to store surplus wind energy from the east. It also increases the security of supply in the event of external disturbances.

Due to the conversion of the transmission grids in the surrounding countries to 380 kV, the remaining 220 kV lines in Austria represent a safety risk. Failures of 380 kV overhead lines in neighboring countries can lead to overloads due to the low capacities of the Austrian 220 kV lines. Grid analyses have shown that building the planned 380 kV lines in Austria will greatly reduce the risk of overloading. The construction of an Austrian 380 kV safety ring from overhead lines would significantly improve fault mitigation. To justify the line expansion, the avoided disturbance costs can be compared to the expansion costs.

The costs of the energy not supplied, according to Table 6.9, can be used as a measure of the economic damage to be expected in the event of a large-scale supply interruption. The mean values of the interruption costs are evaluated with the frequency of the share in Austria and result in an average value of the interruption costs of 8 €/kWh. In the literature there are studies with values of 4 to 20 €/kWh per kWh not delivered. For sensitive industrial companies, these values can be significantly higher.

The costs of interruption are different in the individual voltage levels. The risk R is assessed according to the frequency h of occurrence and the resulting mean damage costs W.

$$R = h \cdot W \tag{6.25}$$

Table 6.9 Costs of undelivered energy in Austria

End use	Share in Austria	Interruption costs	Mean value
		€/kWh	€/kWh
Industry	45%	5–20	10
Household	29%	0.5–5	3
Service	20%	5–35	10
Traffic	3%	10–50	20
Agriculture/forestry	3%	2–5	4
Total	100%		8

Table 6.10 Disturbance costs in the grid levels in Austria

Level	Voltage	Non delivered power	Affected customers	Disturbance per hour
1	380 kV	5000 MW	7,000,000	40 million €/h
3	110 kV	250 MW	250,000	2 million €/h
5	10–30 kV	20 MW	40,000	0.16 million €/h
7	0.4 kV	0.3 MW	300	2400 €/h

Table 6.10 shows the disruption costs for the individual voltage levels in Eastern Austria. If the 380 kV voltage level fails, all subordinate voltage levels are also affected. In case of failure in a 110 kV system, only a few subordinate grids are affected, since these are operated separately from each other 110 kV grid and supply only an average of 250,000 customers. Due to the high impact, the networks for 110 kV and 380 kV have to be designed according to increased security level. This means that all systems components are at least duplicated and must ensure supply in the event of a component fails.

Medium- and low-voltage networks supply a limited number of customers. They represent about 80% of the total network investment. For cost reasons, a redundant and thus reliable and secure supply structure is built here only in medium-sized and larger cities. In rural networks with a low population density, only simple structures without redundancy are often used, since the number of customers affected and the failure costs are low.

The disturbance costs of voltage level 1 in the east of Austria amount to € 40 million per hour (Table 6.10). Simulations have shown that just by expending the section of the 380-kV line from Südburgenland to Kainachtal, the effects of congestion caused by line failures in neighboring countries can be safely prevented. The investment costs for this line amounted to € 120 million and correspond to the damage costs of a large-scale disturbance in eastern Austria lasting three hours. According to Eq. (6.25) an expansion of this line can be justified from the avoided disruption costs.

6.7.3   Energy Economic Analysis of Congestion Management

The transmission capacities of energy grids must be designed to match the development of generation, in particular the more power-oriented demand from renewable energy sources and the increase in network load due to the substitution of fossil energy use. For this purpose, network development plans are used, which annually show the necessary network developments through expansion or network reinforcement in a forward-looking manner over a period of 10 years. If the planned projects cannot be installed on time, but the renewable generation capacities are further expanded, there will be grid congestions. The electricity market can have balancing effects in the case of small congestions and enable efficient use of the existing network capacities.

The divergence between generation and consumption makes extensive market activities and technical congestion management necessary [Christiner 2015]. Grid expansion according to a grid development plan is the measure with the lowest costs and the highest effectiveness.

In the event of severe network overload, the security of supply is endangered and lines can be switched off due to overload. This is often the beginning of a cascading further shutdown up to a blackout.

This dangerous situation can no longer be solved by mechanisms of the free electricity market, but requires the intervention of the system operators. For this purpose, mechanisms for network congestion management such as redispatch, feed-in management and the provision of network reserves have been created.

The following measures can be used [dena 2017], [Bundesnetzagentur 2017], [Agora 2018]:

- **NORE** monitoring: **N**etwork **O**ptimization before **R**einforcement before **E**xpansion. This also includes the use of power-controlling elements such as phase-regulating transformers and FACTS.
- Introduction of wide area overhead line monitoring in order to utilize network reserves dynamically.
- Maintaining grid reserve in order to reduce redispatch and generation management.
- Redispatch of supply according to the available network capacity by throttling or increasing power plant output.
- Feed-in management with the reduction of feed-ins from renewable energy or from CHP plants in highly loaded grid areas. This can be linked to a redispatch with replacement feed-ins from power plants in order to reduce network loads.
- Simplification and acceleration of approval procedures for network expansion.
- Preference for low-cost solutions with high transmission capacities in order to reduce the number of parallel routes and to increase approval acceptance.

The annual costs for redispatch and congestion management through feed-in management result from the replacement remuneration c_{Ai} per MWh not supplied over the shutdown time T_{Ai} for non-delivered energy and the generation costs for substitute feed-ins (index E).

$$K_{EPM} = \sum_i c_{Ai} T_{Ai} + \sum_k c_{Ek} T_{Ek} \qquad (6.26)$$

In case of renewable generation, a distinction must be made between energy, that is transportable but not usable, because it exceeds the load demand and the storage capabilities and energy that is usable but not transportable, because grid capacities are lacking. In the first case, generation capacity must be curtailed by feed-in management. For the second case, compensatory generation is necessary. Other cost

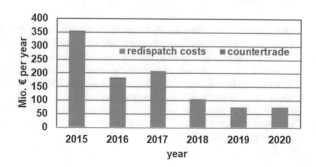

Fig. 6.16 Redispatch and countertrade in the German transmission network [entso-e 2020]

factors of redispatch, that have to be compensated are e.g., lost profits from intraday transactions, lost revenues from avoided network charges and overhead costs to this.

As Fig. 6.16 shows, the redispatch of power plants has a much larger share than the countertrade of renewable energy in generation management. The costs of resolving grid congestions in Germany reach values of the range of € 350 million to € 75 million per year. Redispatch is expensive, since energy is renumerated twice according to Eq. (6.26) and additional cost factors can be added.

The grid development plan in Germany [Bundesnetzagentur 2018] envisages a need for new grid extension of 8850 km by 2030, including 2150 km of new DC – voltage lines.

6.8   Summary

Cables have become established in the medium and low voltage sector. They can be laid cost-effectively in valleys by plowing them in and are well suited for grid operation in terms of their technical and energy-economical characteristics.

Overhead lines are indispensable for high and extra-high voltage. They have high transmission capacities that are well above the nominal values for 35 °C at wind speeds of 0.6 m/s at lower temperatures and higher mean wind speeds.

This makes them particularly suitable for connecting wind farms to the grid. Overhead lines have significantly higher transmission capacities than underground cables. Their reliability is significantly higher than that of cables, due to the short repair times. Due to their low reactive power demand, they are self-compensating and do not require any reactive power compensation measures. They are therefore suitable for large transmission capacities and distances.

High and extra-high voltage cables have a high reactive power, which limits their transmission capacity and range. The long repair times require several parallel cable routes. The investment costs for cables are 5 to 10 times higher than for overhead lines. From a technical and energy economic point of view, overhead lines are therefore the better solution in the transmission network. High and extra-high voltage cables are mainly used in urban supply networks with small areas.

References

[Agora 2018]   Toolbox für Stromnetze. Studie, Agora Energiewende (2019)
[Auckland 1998]   Auckland Power Supply Failure 1998. The Report of the Ministerial Inquiry into the Auckland Power Supply Failure. Ministry of Commerce of New Zealand (July 1998)
[Brauner 2017]   Brauner, G.: Energiewirtschaftliches Gutachten „Errichtung und Betrieb einer Starkstromfreileitung (Ersatzneubau APG-Weinviertelleitung). Gutachten Im Auftrag der Niederösterreichischen Landesregierung (2017)
[Brakelmann 1985] Brakelmann, H.: Belastbarkeit der Energiekabel. VDE Verlag (1985)
[Bundesnetzagentur 2017]   Entwicklung   der   Redispatchmaßnahmen   im   deutschen Übertragungsnetz (2017)
[Bundesnetzagentur 2018]   Netzentwicklungsplan (NEP) 2017–2030

[Christiner 2015]   Christiner,   G.:   Windenergie   und   Herausforderungen   für   die
Übertragungsnetze. WEC-Workshop Windenergie auf dem Weg zur Grundlast. Wien
6.10.2016

[dena 2017]   Höhere   Auslastungen   des   Stromnetzes.   Massnahmen   zur   höheren
Auslastung des Bestandnetzes durch heute verfügbare Technologien. Deutsche Energie-
Agentur Gmbh (dena) (2017)

[e-control 2016]   Statistik: Trassen- und Systemlängen zum 31.12.2016

[entso-e 2020]   entso-e: Costs of Congestion Management (2020)

[Fischer 1989]   Fischer, R., Kießling, F.: Freileitungen—Planung, Berechnung. Springer,
Ausführung (1989)

[Hütte 1988]   Elektrische Energietechnik, vol. 3. Springer, Netze (1988)

[Jupe 2011]   Jupe, S., Bartlett, M., Jackson, K.: Dynamic thermal ratings: The state of
the art. 21st International Conference on Electricity Distribution, CIRED, Frankfurt, 6–9 June
2011

[Kabel-HB 1986]   Kabelhandbuch, 4. Aufl. VDEW (1986)

[Kiwit 1985]   Kiwit,   W.,   Wanser,   G.,   Laarmann,   H.:   Hochspannungs-   und
Hochleistungskabel. VWEW (1985)

[Niemeyer 1992]   Niemeyer, P.: Freileitungen. VDE (1989)

[Oeding 2004]   Oeding, D., Oswald B.: Elektrische Kraftwerke und Netze, 6. Aufl. Springer
(2004)

[Oswald 2007]   Oswald, B.: 380-kV-Salzburgleitung – Auswirkungen der möglichen (Teil-)
Verkabelung des Abschnitts Tauern-Salzach neu. Gutachten im Auftrag von Energie-Control
GmbH Wien (2007)

[Oswald 2010]   Oswald, B., Hofmann, L.: Gutachten zum Vergleich Erdkabel—Freileitung
im 110-kV-Hochspannungsbereich

[Siemens 1987]   Kabel und Leitungen für Starkstrom. Teil 1 und 2. 4. Auflage 1987. Siemens
(1987)

[Sinowatz 2015]   Sinowatz, P.: Rechnerisch „stromautark" mit Windenergie. WEC-
Workshop: Windenergie auf dem Weg zur Grundlast. Wien 6.10.2015

Efficiency of the Decentralized Energy Supply

7

7.1 Challenges of Sustainable Energy Supply

The integration of renewable energy sources into the distribution grids transforms them from passive to active networks. Customers convert from passive consumers to producers and consumers, so-called prosumers. Since the renewable energy sources for the provision of the final energy demand have high power with low full load hours, the unplanned integration of the renewable energy can lead to a significant grid expansion with a short usage period and thus to inefficient grids.

The current models with exclusive promotion of renewable energy sources and the compulsion to provide grid capacities by the grid operator free of charge, according to the promoted power of the sources, supports these tendencies.

Several approaches can help to solve the new challenges:

- From the point of view of the regulation of energy systems, the subsidy models should be replaced by market models in the medium term. The deregulation of the electricity market should also be maintained in the future. From the perspective of the distribution system operators, this means they cannot intervene in the generation market for capacity management. This is only permitted for congestion management to avert major disturbances. From the regulatory point of view, grid capacity and energy volumes should continue to be the distinguishing criteria for grid and market [BNA 2011], referred as to Smart Grid (Micro Grid) and Smart Market.
- The grid operators see the smart grid with the control of the final consumption according to congestion criteria of the grids as a possible solution. From the regulatory point of view, efficient grids should remain in place even if generation is predominantly renewable. Therefore, grid expansion should not be designed to transport any peak load, but the market should help to eliminate grid bottlenecks [BNA 2011].

© Springer Fachmedien Wiesbaden GmbH, part of Springer Nature 2022
G. Brauner, *System Efficiency by Renewable Electricity*,
https://doi.org/10.1007/978-3-658-35138-0_7

- The Micro Grid or the decentralized energy cells are an approach to coordinate generation close to the consumer and local storage in such a way that the need for grid expanded is kept to a minimum and long-distance transports of short-term generation peaks are avoided.

In general, the following criteria will continue to apply to sustainable energy supply in the future:

- The energy supply should be largely sustainable.
- The prosumers should be able to easily deliver and charge their excess energy to other consumers in the distribution grid.
- The intelligent grid or micro grid will be equipped with smart meters and information and communication technology (ICT). Data security must be secured here. Furthermore, in the event of an ICT failure, a limited full supply to end consumers should be possible, with no overloading of operating equipment and no significant violation of the permissible voltage bands. It should also be possible to restart the network quickly and not depend on a longer-term restart of the information infrastructure.
- The ICT infrastructures required for supply security and network restart should be independent in their self-supply form an energy supply from the networks to be controlled.
- The renewable supply should be as reliable as before.

7.2   Decentralized Supply Concepts

7.2.1   Smart Grid

Examples of research activities in the field of Smart Grid are Web2Energy 2012 in Germany and Europe and the DG DemoNet 2015 in Austria. With a high proportion of photovoltaics, decentralized small generators and electromobility in the distribution grid, the research projects are investigating suitable methods of intelligent planning, monitoring and grid control for secure grid operation.

The smart meter is the essential component for monitoring [DG DemoNet 2015, Web2Energy 2012]. For grid control, controllable distribution transformers, controllable loads especially as charging stations for electromobility and as electric heating systems with heat pumps, and low-voltage string controllers are used to improve voltage stability in low voltage feeders.

The introduction of market models as Smart Aggregation [Web2Energy 2012] or as a Smart Market [BNA 2011] represent a supplement to the more technically oriented Smart Grid concepts and allow the basic principles of deregulation to be retained in the Smart Grid. For example, controllable loads can be controlled from the perspective of technical grid management or via incentive systems in the smart market. The use of storage capacities in the distribution grid can also be seen as regulatory congestion management or as free energy trading. The organizational separation of grid and market in the context of deregulation will define the regulatory boundary lines between smart grid and smart market in the future.

The developments are not yet completed. They are to be understood as an evolutionary process in which partners from the fields of energy supply, energy regulation, energy technology and the energy market will participate in future with new concepts and strategies.

7.2.2   Micro Grid and Decentralized Energy Cells

The Micro Grid represents a cellular approach [BNA 2011], in which energy cells generate and store their own energy demand with the highest possible proportion, and use flexible loads to align their demand more closely to the renewable supply. The aim here is to develop efficient decentralized supply structures through demand-oriented planning of decentralized capacities. At the same time, the necessary expansion of the superimposed transmission grid should be made possible for longer time of use (full load hours) and thus for greater economic efficiency.

The concept of the Micro Grid provides for a high degree of independence and autonomy for the energy cells. The cells can determine their own generation and control concepts. The exchange of information between the cells and the superimposed infrastructures should be kept at a low level. In the event of failure of the superordinated grid, it should be possible to ensure a temporary emergency supply of the decentralized cells until the central security-relevant systems are rebuilt.

The Micro Grid concept requires the following solution strategies:

- The decentralized generation capacities from photovoltaics and other small power plants are to be designed in such a way that they enable the highest possible coverage rate for cellular demand.
- The transmission capacities of the cellular distribution networks to the higher-level grid should not be overloaded by exporting excess energy generated from renewable sources. Instead, local storage capacities at the end users should store the surplus of peak power and shift them to times of low generation.
- The end customers of the cells should have the possibility to exchange energy with each other as micro-balance groups to increase the overall utilization rate of the cell. In large buildings in particular, owners and non-owners should be able to use PV as fully as possible with automatic decentralized billing systems (tenant electricity model).
- The decentralized generation units, such as photovoltaic systems, and the end use of significant power, for example, in charging stations for electric cars, should be distributed as evenly as possible across the grid. In this way, the string loads can be reduced and the voltage maintenance in all grid areas can be kept in the same direction, with enables grid control with controllable local grid transformers.

- The incentive financing of renewable generation facilities should consider the grid loading and only support facilities that have a high degree of self-use of the energy generated and thus have a minimal need for grid capacity.
- Trough framework conditions such as the building code, PV systems with corresponding storage units should be prescribed for new construction projects and, in doing so, the band of PV power to be installed and the storage capacities per residential unit should be specified. From the point of view of cellular distribution grids, this also results in fixed rules for the grid connection capacity to be provided per residential unit.
- Diversity and independence of the control and information systems of the decentralized cells should result in high security against wide-area attacks from the Internet.

In the following, the concept of the Micro Grid is considered and an efficient design for the integration of photovoltaics and electromobility is examined through time series simulation.

7.2.3   Vertical Grid Usage and Traditional Electricity Market

The distribution networks are traditionally used vertically. Here, the end users in the low-voltage network do not have their own generation facilities. The facilities for generating electricity are in the higher-level networks. The end customers purchase their energy from energy suppliers or from other energy service providers. The electricity market is organized in balance groups. The price component for the energy is about one third of the total tariff. The share for grid usage is another third and the remaining third consists of taxes and levies.

Figure 7.1 shows the structure of vertical grid utilization. In the case of wind energy, the renewable energy sources are grouped into wind farms.

Fig. 7.1 Vertical grid usage with renewable generation

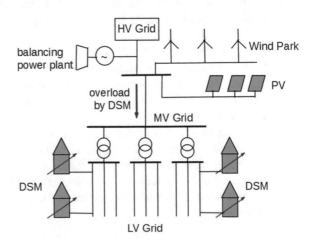

The photovoltaic systems are realized in free field setups with high outputs. Furthermore, storage units can be realized as pumped storage units or central accumulator plants in the grids. The structure of this grid usage largely corresponds to the traditional generation structure with large power plants, except that these are replaced here by renewable generation plants in large-scale setups. The thermal power plants are still required to bridge periods with low supply of wind energy and photovoltaics with balancing energy.

In the traditional fossil energy supply with power plants, the output of the power plants is adjusted according to the final energy demand, and the need for balancing energy for temporary technical differences between generation and demand can be met with a few pumped storage facilities.

When changing over to renewable energy sources, the supply determines the generation characteristic and the load characteristic of the demand can deviate significantly from it. Therefore, significantly larger storage capacities are required and power plant reserves in the order of magnitude of the peak demand must be kept available.

Table 7.1 shows the different characteristics of generation and demand. The generation factor indicates the factor by which the power of the generating plant must be greater than the peak load so that the same annual energy can be generated according to the load profile H0.

The annual energy demand of the load E_L is calculated from the peak power of the load P_L and the full load hours T_L

$$E_L = P_L \cdot T_L \qquad (7.1)$$

The renewable generation plants must install higher capacities in accordance with their lower full load hours T_i.

$$E_i \equiv E_L = P_i \cdot T_i \qquad (7.2)$$

$$P_i = P_L \cdot \frac{T_L}{T_i} = P_L \cdot p_i \qquad (7.3)$$

For a full supply from photovoltaics, 3.9 to 5.2 times the peak power of the grid load would have to be installed according to the generation factor p_i, if storage of any size is available. In reality, even larger generation capacities are required

Table 7.1 Full load hours and generation factor of generating plants

	Full load hours T in h/a	Generation factor p_i
Load profile H0	4680	1.0
Photovoltaics	900–1 200	3.9–5.2
Conventional wind energy onshore	1800–2 200	2.1–2.6
Low wind turbine onshore	2500–3 800	1.2–1.9
Wind turbine offshore	3500–4 800	0.98–1.4
Run-of-river power plant	4500–5 600	0.83–1.04

to mitigate the storage demand. In the case of a central installation, high network transmission capacities must be provided that are only used for a short time. The PV peak load only occurs for a few hundred hours, which makes such grids uneconomical. Furthermore, very large storage capacities would have to be created to shift the overproduction in summer to the winter half year or vice versa for wind energy. This is neither economically nor technically expedient. This means that large centrally installed PV systems are not very suitable for full supply. In Austria, these systems have no longer been subsidized since 2015.

The low-voltage grids are not overloaded in the vertical grid use, provided that the loads do not grow, or the load characteristics are not changed, e.g., by spread and intensive use of fast-charging facilities for electric vehicles.

To avoid the expansion of large storage capacities or the curtailment of renewable generation sources in the event of overproduction, the demand of loads can be made flexible. In the deregulated electricity market, however, this is not permitted to the grid operator, but only to the electricity market. For this purpose, the end customer must be given economic incentives. Renewable plants have only fixed costs and no variable costs (fuel and operating costs). The marginal cost of generation is therefore near zero. Even if the renewable energy is offered at tariffs of zero, this only means a tariff reduction of a third for the end customer, since the grid costs and taxes as well as levies remain unchanged. Therefore, the economic incentives for supply-oriented demand side management are not very high. With intensive use of a supply-oriented demand side management, grid overloads are possible.

Centralized large-scale PV plants (Fig. 7.2) have future risks in the areas of market, grid, and the environment. It is likely that today's support strategies for renewable energy will be replaced by market models. Since solar energy is

Fig. 7.2 Central PV system in free field installation

determined by large-scale weather patterns, on clear days many such plants are active at the same time and there is an oversupply on the stock exchanges, causing the market price to move towards marginal costs, i.e., towards zero. These centralized plants are therefore only economical with today's funding mechanisms.

Environmental regulations pose further risks. Due to the consumption of landscape in the case of large-scale installation and the competition with use as agricultural or forest land, there will be further risks here in the future. Finally, the need for high grid capacities for connection and transport with only short usage time, represents further economic risks in these grids, as costs are no longer socialized but are assigned to the plant operator.

7.2.4 Horizontal Grid Usage and Smart Market and Micro Market

New concepts of decentralized energy cells or the micro grid try to avoid the disadvantages of centralized generation and vertical use. Here, the renewable generation is to be located close to the place of electricity demand, to avoid long-distance transits at high power levels. Figure 7.3 shows such a concept. Photovoltaic systems are only installed on buildings. There, they are equipped with local stationary accumulators. The power of the photovoltaic systems and the storage are designed in such a way that the daily demand of the building can be covered in summer, but significant power flows into the low-voltage grid are avoided. Therefore, from the perspective of the overlaying grids, the PV systems act as load reducers but not as generators, and no grid expansion is required for this purpose.

In the individual local medium-voltage grids low-wind turbines are installed, whose power is limited by smaller generators and whose full-load hours are

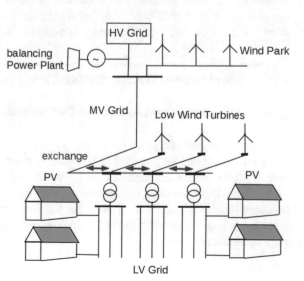

Fig. 7.3 Horizontal grid usage with renewable energy supply

increased by larger rotors. They therefore do not require a significant power-oriented grid expansion and have a reduced need for storage. The integration of these systems in the medium-voltage grid allows surplus energy to be used by neighboring local grids.

By appropriate planning of the PV and wind energy plants as well as storage facilities, annual local renewable coverage rates of 80% can be achieved, as shown in the following simulations with time series of PV and wind generation and with synthetic load profiles. There is then only an exchange of balancing energy between the small-scale energy cells. It is true, that the distribution grids are still to be designed in such a way that the full supply is possible from the superimposed grid in periods without significant wind and PV potential. However, an oversizing of the grids for short-term use of renewable generative plants with high output power at short full load hours is avoided.

The distributed renewable generation plants can also be used to maintain the voltage of the distribution network. This increases the available network capacity. The capacity of a line determined by its kilometric impedance X'_L, the line length l and the voltage control at the beginning of U_1 and at the end of U_2, as well as by the load angle δ between input and output voltage.

$$P = \frac{U_1 \cdot U_2}{X'_L \cdot l} \sin\delta \tag{7.4}$$

In passive vertically used grids without self-generation, voltage sags at the end of the line during high load are a problem, that can be remedied with distributed generation at the end.

Generation balancing between the individual micro grids can be remunerated by a smart market according to cause. Here, aggregators trade surplus energy and try to achieve a small-scale energy exchange to reduce network costs. The remote reading of data from smart meters is gaining in importance as an indicator for trading options. Within buildings, automated electricity trading systems are also possible, which can offset local PV use by households in large buildings.

7.3 Efficiency of Decentralized Energy Cells

7.3.1 Efficiency Strategy of the Decentralized Energy Supply

Coverage ratio over the year

Wind energy can be installed in rural regions with a safe distance from settlements. Photovoltaics are installed on the roofs and facades of the buildings. The grid infrastructure consists of low and medium voltage grids, which have limited transmission capacity. As has already been shown, the expansion of the electricity grids according to the peak power of the installed renewable energy sources is not very economical.

The concept of decentralized energy supply is therefore as follows:

- The energy output and the generation characteristics of the installed renewable energy sources are to be matched to the demand of the decentralized energy cell in such a way that as large a proportion as possible can be used locally.
- The electricity grids are required for the transport of the balancing energy and should be left in their expanded state as far as possible. The existing network capacities therefore determine the highest possible balancing power.
- The highest possible degree of utilization of the locally available energy sources requires a concept for the internal exchange between the prosumers of an energy cell, i.e., within large buildings with significant PV installations, cost allocation based on causation should be possible between PV owners and PV users. Within distribution grids, simplified cost-justified exchange should also be possible.
- Surplus energy should also be usable locally via tariff incentives through sector coupling for electromobility, for decentralized electricity storage, heating, or cooling systems with heat pumps or also for electrical storage heating. The coupling of the sectors for electricity and heating or cooling represents great potential for balancing the energy cells.

7.3.2 Possible Coverage Rate of Decentralized Energy Supply

Synthetic load profile method
First, a local supply from wind energy and PV without storage and load management should be considered. Since supply concepts using isolated small cells are not very economical due to the high possible peak load with a short period of use, larger grid connected supply collectives are to be considered. For this purpose, the method of the synthetic standardized load profiles is used [BDEW 2008]. Here, the individual consumer groups are assigned load profiles based on year-round measurements of the quarter-hourly consumption. The energy service providers feed the respective energy service into the grid according to the number of customers for a profile.

Figure 7.4 shows the load profiles for households in winter (H0 winter load profiles) normalized to an annual consumption of 1 MWh. Within a week, special

Fig. 7.4 Standardized H0 load profile for households in winter [BDEW 2008]

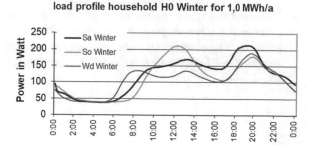

load profile household H0 Winter for 1,0 MWh/a

profiles are used for the weekdays (Wd), Saturday (Sa) and Sunday (So). The winter profiles apply from November 1 to March 20. Further profiles apply for the transition period from March 21 to May 14 and from September 15 to October 31. Likewise, summer load profiles are provided for the summer period from May 15 to September 14.

There are also standardized load profiles for small businesses (bakers, hairdressers, stores, offices). All load profiles are normalized to an annual consumption of one MWh and are converted according to the actual annual demand. Based on measurements, many energy service providers use individualized load profiles, especially for large customers. In the following, the H0 load profiles of all annual periods are used to analyze the coverage rate in residential areas over the year.

The synthetic load profile method can also be used to determine the possible coverage rates for renewable energy supply.

Coverage rate without local storage and load management
The following degrees of freedom exist when determining the possible coverage rate for renewable generation:

- The installed wind capacity in the onshore area per end customer and the generation characteristics of the wind turbine as normal or low wind systems can be selected. A low wind turbine with a generator output of 3 MW and a rotor output of 6 MW is selected, as it enables a high coverage rate, due to the high full load hours.
- The installed PV power per end customer represents a further degree of freedom. Since it is assumed that PV systems are only installed on buildings, their output must be matched to the end demand and should not cause any overload in the low-voltage grid with simultaneous generation on clear summer days. Due to the design criteria for low-voltage grids with a minimum connection power of 2 to 5 kW per end user, load of 3 kW per household is selected. Considering an average household peak load of 1 kW, the grid load is thus limited to about 2 kW each with simultaneous feed-back from all households.
- The mixing ratio of wind energy and photovoltaics is another degree of freedom. Previous studies have already shown that the most favorable mixing ratio of wind energy to photovoltaics in inland areas is about 3:1 [ADRES 2009]. This ratio is for annual energy delivered and not to the installed generation capacity. With a low-wind system output of 3 MW and the average full load hours of PV systems in Central Europe, this ratio is established at 3 MW PV.
- The number of households to be supplied represents the third degree of freedom of the necessary over-generation. First, the coverage rate is examined based on a time series analysis with generation time series of PV and wind and standardized household load profiles. Initially, no local storage facilities are considered.

Fig. 7.5 Daily generation from wind and PV

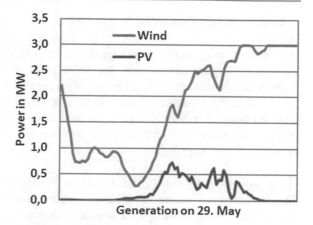

Generation on 29. May

Fig. 7.6 Generation capacity and coverage rate with a low-wind system

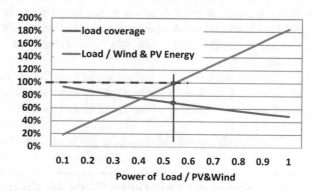

Power of Load / PV&Wind

Figure 7.5 shows a section of a time series of a week with simultaneous generation from wind energy and photovoltaics for installed capacities of 3 MW each. Here the wind generation reaches the full generation output.

With an assumed PV power of 3 kW per household, the number of households to be supplied is predetermined. However, it should still be considered as a degree of freedom.

Figure 7.6 shows the result of the time series analysis as possible annual coverage of the household load. The values of abscissa axis represent the ratio of the total load of all households to peak load of wind and PV generation. The value 0.5 means that the total household load is 3 MW with installed wind and PV power of 3 MW each, together 6 MW.

The red curve represents the annual generation from wind and PV in relation to the annual energy demand of the load. The value 100% means that the annual generation from wind and PV corresponds to the annual demand of the load. Due to the different time profiles of generation and load, here only 70% of the generated energy can be used (blue curve). The power of the load (abscissa) is 55% of the total installed power from wind and PV.

Figure 7.6 also shows that a high load coverage rate can be achieved at low load. However, only a small part of the generated renewable energy can be used. Furthermore, the annual energy demand of the load should be lower in relation to the renewable generation (below the 100% point of the red curve) to enable the highest possible coverage rate at a low power export via the grid.

The abscissa value of 0.5 represents a practical design. This means that the installed nominal power of light wind systems and PV should be approximately equal and together should correspond to twice the peak load of the region with households to be supplied.

Coverage rate with storage and local load management
The decentralized utilization rate of locally generated renewable energy can be increased through storage capacities or sector coupling of electricity to heat. The provision of heat by heat-pumps or electric storage heaters enables significant potentials, with peak power in the range of up to about five times the value of the electricity needs of households. The outdoor temperature determines the applicability, making this possible predominantly in the winter months as load management.

Accumulators from PV systems can also absorb surplus energy. This is possible with PV mainly in the summer months as well as throughout the year with wind energy. Controlled charging of electric vehicles is another option for load management.

Table 7.2 shows possible potential per household.

7.4   Generation Efficiency Through Coupling of Regions

Renewable energy from wind and solar radiation is subject to fluctuations in supply. In the case of photovoltaics, passing cloud fields can modulate the solar supply. Wind energy is created by balancing air flow between high- and low-pressure areas. In high-pressure areas, the global radiation increases the air pressure. Starting from the center of the high-pressure area, a flow in all directions is generated. There is no wind in the center itself. In the low-pressure areas, the air flows in from all sides. Also, here in the center the wind speed is zero. The high and low-pressure areas themselves have a migration speed of around 20 to 30 km/h.

Table 7.2 Potential for storage in households

Application type	Connected load	Annual energy demand
PV with accumulator	1–5 kW	500–5000 kWh
Heat pump	2–4 kW	1000–5000 kWh
Electric storage heater	2–10 kW	1000–10000 kWh
Electric vehicle	2–50 kW	1000–2000 kWh

Fig. 7.7 Spatial interconnection of low wind systems

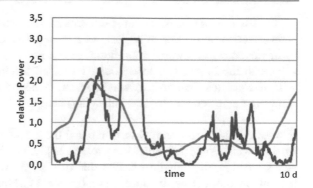

Fig. 7.8 Annual duration curves of wind turbines with spatial networking (NW normal wind, LW low wind, PR rotor power, PG generator power)

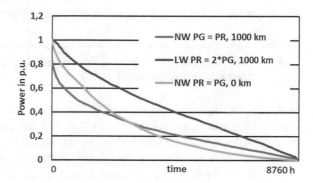

The spatial interconnection of wind energy and PV systems enables averaging of supply fluctuations. The interconnection of energy regions via electricity grids is therefore advantageous and enables the exchange of balancing energy to reduce fluctuations in supply.

The time series of the wind supply of Lower Austria and for a migration speed of the wind supply areas of 20 km/h, the annual duration lines were determined for a normal wind turbine and a low wind system.

Figure 7.7 shows the effect of a spatial interconnection of low wind systems over an area of 1000 km at a weather front migration speed of the of 20 km/h for a period of 10 days. The large gradients in local wind generation (red) are replaced by lower global ones and the short-term gradients of peak power are replaced by a more even power curve (blue). From the point of view of grid operation, this also reduces the balancing energy required and makes it easier to plan.

As Fig. 7.8 shows for a normal wind turbine, that spatial interconnection results in the following changes in the annual power-duration curve:

- With a single normal wind turbine as a reference, the peak output reaches the nominal output, even if only briefly (green curve).
- If the plants are distributed over a distance of 1000 km (blue curve) and the total of the individual system outputs is defined as the nominal output, the

result is a lower peak output. Furthermore, the annual duration curve changes in direction of a lower slope, i.e., for a more uniform generation. This means an improvement in towards a minimum generation.

* Due to the higher full load hours, low wind turbines (red curve) have better characteristics than normal wind systems (blue curve), especially in the lower power range and require less balancing energy.

If wind turbines are installed in a line and the spatial propagation of the weather fronts travels along this line, these results apply. If the weather fronts run transversely to this, the situation is like that of a single wind turbine. However, this has only theoretical significance. In reality, wind turbines are installed over a wide area. If these turbines are also interconnected over a wide area, favorable conditions arise for compensation through spatial interconnection.

7.5   Self-Generation and Coverage Rate in the Distribution Grid with PV

In the low-voltage distribution grid, in the future, photovoltaics will be the most common renewable generation technology. For this purpose, the installed power of photovoltaics must be matched to the demand. It should also be checked which improvements in energy use are possible through decentralized accumulators. Finally, a cost–benefit analysis must be carried out, to enable an efficient use of the investments.

In the future, photovoltaics will be installed almost exclusively on buildings. From the point of view of the owners, the profitability results from the fixed costs for depreciation of the investments for the delivered annual generation, compared to the electricity purchase tariffs from the grid (grid parity).

PV systems without storage are already competitive compared to the grid tariff. In future, systems with storage will also be economical. The layout of the plants will mainly be for economic reasons. Owners of these systems will attempt to use as high a proportion of the PV energy as possible themselves, since export via the grid in the future subsidy-free renewable market is not economical, due to the low marginal costs of generation. Therefore, the following sections examines PV systems design that have the lowest possible investment costs while meeting the owners' own needs at the highest possible rates.

From the perspective of a holistic view of the distribution network, load profiles can be used for the individual end users. Likewise, the PV systems and the accumulators should be evenly distributed to all end users. This also represents an economical solution from the perspective of the end users since each end customer then sees the fixed costs of their decentralized installation in relation to the purchase tariffs from the grid.

If individual end users have larger installation capacities, they can only partially use the energy gained from this and must export the surplus into the network. The end users with no installation or one that is too small then must obtain their demand from the grid at the valid grid tariffs, which eliminates the benefits of lower PV costs for their own use.

A one-year series of measurements in Lower Austria is used as the time series for the solar supply. Measurements were carried out synchronously at fourteen locations spread across Lower Austria, and an average value was calculated from this. This time series of mean values is used for the analysis below.

The H0 load profile for households is used for end customers, which has a peak load of 277 W per end customer with an annual requirement of 1 MWh. This is extrapolated to the total load, as is common when using synthetic load profiles. First, the use of PV without accumulators is considered. The PV systems are only installed on buildings, which means that they can be used cost-effectively without transport via the distribution network.

Figure 7.9 shows the possible coverage rate of the household load. The ratio of the installed PV power to the average peak load of the households is varied according to the H0 load profile. Up to a ratio of two of the PV outputs to the load, the degree of coverage rises sharply and flattens to higher power ratios. This means that due to the PV generation profile and the load profile, load coverage of up to 35% is economically possible. The reason for this is that solar radiation is only available during a short period of the day and without storage no use is possible during the rest of the time.

Figure 7.10 shows the PV energy (red) and the coverage rate of the average household load (blue). If the PV output is more than a factor of 2 compared to the peak load of the household, the coverage rate only increases slightly. The excess PV power must be offered in the future in the free electricity market at marginal cost of PV generation close to zero.

The costs for PV installations grow linearly with power, the gradient of the grows of the coverage rate becomes smaller and smaller as the PV power increases, making the installation of large power uneconomical. Therefore, from the perspective of PV systems owners, a system power of two to three times the peak load of the load profile is therefore economical.

If electric vehicles are to be supplied with power, a correspondingly modified load profile with a higher peak load must be used.

Fig. 7.9 PV coverage rate

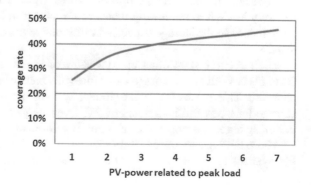

Fig. 7.10 Comparison of PV
generation and coverage rate
of the household load

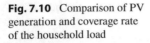
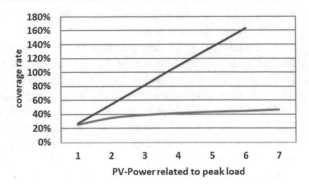

PV-Power related to peak load

Fig. 7.11 Coverage rate with
solar storage systems

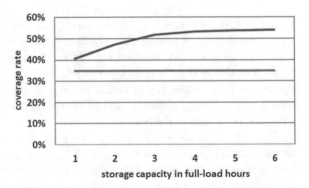

storage capacity in full-load hours

From the point of view of the distribution grid, the available transport capacity
per household is only about two to three times the highest average peak load. It is
therefore advisable to limit the PV power to peak load to a factor of two.

For the load ratio of two, which is favorable from the point of view of the distribution grid, it is now again investigated by means of time series analysis which
improvements in the load coverage rate are possible through the use of local storage capacities.

Figure 7.11 shows the shares of direct coverage without storage (blue) and the
improvement in the coverage rate by storage (red). The ratio of the local storage
capacity in kWh to the locally installed PV power in kW is varied. Formally, this
corresponds to the storage capacity in full load hours of the PV installation on the
abscissa.

The red curve shows that up to a value of 3, the coverage rate increases from
35 to 50%. With larger storage capacities, this hardly brings a higher coverage rate
anymore. This means that solar storage systems are only economical as daytime
storage for a few full load hours of PV. Over the year, larger storage capacities do
not bring higher coverage rates and are therefore not economical.

These findings are used in Table 7.3 for a compilation of the planning design of
PV systems for different applications areas. It is assumed that there is an annual

Table 7.3 Planning values for PV systems with solar accumulators (RU: residential units)

	RU	People	Peak load	PV power	Roof area	Accumulator
			kW	kW	m^2	kWh
Detached house	1	2	1.1	2	15	6
Apartment building	10	15	8.5	17	100	50
Large building	100	150	55	110	650	330

electricity requirement of 2 MWh per person and that there are on average 1.5 persons per residential unit.

Electric heating systems are not considered in the load when planning PV systems, since the main heating demand is in winter, but solar radiation occurs mainly in summer. Systems for room air conditioning, on the other hand, must be considered.

The time series analysis shows that by the extensive use of decentralized PV with local storage can cause utilities to lose up to 50% of their electricity sales in the residential sector. However, since about 50% of decentralized PV energy is fed back into the grid, new business models are possible for utilities to aggregate this energy and trade it, which can at least partially compensate for these losses. In the distribution grid, the energy suppliers are thus converting from being pure energy supplier to energy traders as well.

7.6   Sustainable Electric Mobility in the Distribution Grid

Charging profiles and location distribution of electric vehicles
The conversion of cars to electric drive, and from fossil fuels to sustainable generated electricity from wind, solar, hydropower or biomass represents an important pillar for low-emission and efficient mobility. The mobility sectors can be broadly divided into:

- Inner-city traffic. Local public transport dominates here with subway, tram, and electric bus.
- Rural and suburban local transport. This will be a major sector of individual transport in the future. In this application sector, which accounts for most of the private traffic, a fully renewable and emission-free mobility is possible in the future.
- Long-distance transport by rail. Here, electrification is already well advanced.
- Long-distance traffic on roads and highways. Drive technology here is still largely dominated by the internal combustion engine. Because of the high speeds, the specific energy demand is high, and the range of electric vehicles is therefore much shorter than in local transport. Fast-charging stations are

Fig. 7.12 Normalized load profile for the loading at home [Leitinger 2011a]

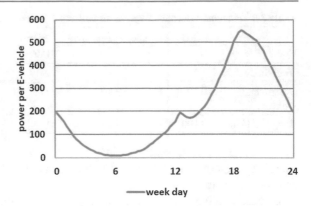

Fig. 7.13 Standardized charging profiles for charging at home and the place of work [Leitinger 2011b]

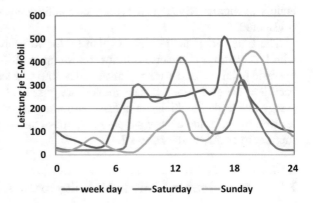

required for energy supply, which must be supplied directly from medium- or high-voltage grid due to the high power levels.

In the long-distance transport sector, in addition to electric drives with accumulators, vehicles with fuel cells and hydrogen storage systems will also be possible in the future. Like vehicles with internal combustion engines, these will then have short refueling times and long ranges. This sector will not be considered here.

In rural and suburban local transport, the energy demand is determined by mobility needs. A detailed analysis was carried out by Herry [Herry 2008] in Lower Austria based on a survey. The results of this survey were used by Leitinger in the research project SEM [SEM 2011] and a dissertation to analyze the energy demand for electric mobility. From this, synthetic load profiles were developed (Figs. 7.12 and 7.13). In addition to the mobility demand, the distribution functions of the parking locations for charging of the electric vehicles must also be considered.

The location distribution shows that only about 5% of the vehicles are driving at the same time. All other vehicles are parked in different locations. At night, 90%

of the vehicles are parked at the place of residence, one third of them in a private garage, at a private parking space or in public spaces on the roadside.

During the day, around 40% of the vehicles remain in their parking position at the place of residence. Up to 30% are parked at the workplace and the rest are parked at the place of shopping, drop-off or pick-up or other locations. Based on the parking location distributions, the home location or work location are particularly suitable for electric vehicle charging because of the longer parking durations.

The standing times are short at the other locations. Because of the short distances in local traffic, recharging of the vehicles is usually not necessary at these locations.

Figure 7.12 shows the standardized load profile for charging an electric vehicle at home. The vehicles return in the evening and are recharged.

Since all electric vehicles return home in the same time window, there is an unfavorable charging peak in the evening, which occurs at the same time as the evening load peak in the standardized H0 load profile of households. Furthermore, solar energy from PV systems can only be partially used in the evening.

More favorable are charging strategies that use both the work and the home location. These profiles are shown in Fig. 7.13 for the work week and for Saturday and Sunday. This charging strategy is better suited for solar powered mobility. On a weekday and on Saturday, the course represents a favorable constellation for the consumption of charging electricity from photovoltaics.

The charging profiles shown here apply to a daily energy requirement of 4.9 kWh on a weekday, 3.5 kWh on Saturday and 3.3 kWh on Sunday. This corresponds to driving distances of about 30 km on a weekday, 22 km on Saturday and 21 km on Sunday. An average energy requirement of 16 kWh/100 km was assumed for the vehicles, which includes the energy of the auxiliary units in the vehicle for heating in winter, cooling in summer for lighting and other applications.

Solar mobility
Solar powered mobility is an interesting future application of photovoltaics. An electric vehicle only needs around 1600 kWh per year for an annual distance of 10,000 km in suburban and rural traffic. This corresponds to an installed PV power of around 1.6 kW per electric vehicle and thus a solar area of around 10 m^2 [Brauner 2008]. Due to the different distribution of the solar supply over the year, too much is produced with this area in summer and too little in winter. These values are only an average. An oversizing of PV systems is recommended, so that the solar coverage rate is improved over the year and the more expensive purchase of electricity from the grid is reduced.

By time series analysis with the solar supply, the possible coverage rate or the charging can be determined as an annual average value. The installed PV power is varied in comparison to the peak power from the average charging profile for both charging strategies, According to Figs. 7.12 and 7.13, this is 550 W and 510 W respectively.

With a mixed charging strategy "at home and work", the coverage rate at three times the PV power compared to the peak power of the load profile already reaches a coverage rate of 43%. This corresponds to a PV power of 1.6 kW. This clearly shows that the annual energy of a PV system, which corresponds to the driving energy of one year, can only be partially used due to the different curves of the generation and charging profile.

An oversizing to a generation factor of 7 (Fig. 7.14) allows only a small increase to 50% and is therefore not economical.

The charging strategy "only at home" leads to a low coverage rate of 25% at the more economical generation factor of 3. By using stationary accumulators at the place of residence, the coverage rate can be further increased in both charging strategies. However, this is associated with additional costs for the stationary storage units.

7.7   Sustainable Supply Strategy in the Distribution Grid

For the renewable potential in Eastern Austria, a decentralized supply concept in the micro grid with the highest possible self-generation rate is being investigated. The planning is based on the following key data:

- The average annual energy requirement per residential unit is to be 2 MWh.
- Each residential unit has an electric vehicle that is used around 10,000 km per year.
- The energy supply is to be provided by wind energy and photovoltaics. The optimal energy mix ratio in Eastern Austria of 3:1 should be maintained.
- The photovoltaic systems should be equipped with stationary accumulators, the storage capacity of which should be one to three hours in relation to the photovoltaic peak power.
- The integration of photovoltaics in the settlement is to take place exclusively on the roofs and facades of the existing buildings. The power of the photovoltaics is to be limited according to the demand in such a way that overloading of the distribution grid is avoided.

Fig. 7.14 Annual coverage rate of solar mobility

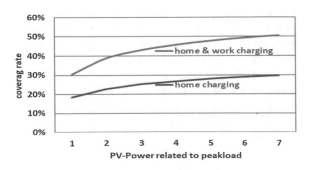

Table 7.4 Load coverage rate for different storage capacities

Storage capacity	Load coverage rate
kWh/RU	
0.0	74.1%
2.0	76.0%
4.0	81.6%

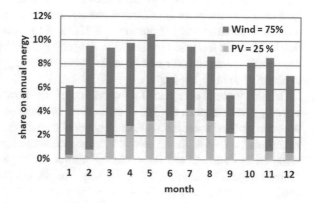

Fig. 7.15 Monthly energy shares from wind energy and PV

For the residential units (RU), the peak load according to the H0 load profile is 554 W with an annual electricity demand of 2 MWh. The car with electric drive is charged according to the "home and work" charging profile and has a peak load of 510 W with an annual energy requirement of 1.6 MWh.

For the sustainable generation of electricity, a proportional output of 2 kW from low-wind turbines is installed per household with an electric vehicle and a proportional PV output of 2 kW is installed on roof areas with optimal alignment. For the generation profiles in Eastern Austria, this results in power shares of 75% for wind energy and 25% for photovoltaics.

Stationary accumulators in the buildings can be used to improve the coverage rate from wind and photovoltaics. Table 7.4 shows the coverage rates for different storage capacities.

By limiting the PV power to 2 kW per RU, an overloading of the distribution grid due to back-feeding during strong solar irradiation in summer is not possible.

Figure 7.15 shows the monthly energy shares from wind energy and photovoltaic for the above example, based on the total amount of energy generated annually. PV generation has the largest shares in summer. Wind energy is complementary to this with the largest shares in winter.

If the supply of heating energy for heat pumps and storage heating is to be considered in the planning, the installed wind power must be increased accordingly for this purpose. Due to its annual distribution in the winter half-year, PV can only contribute slightly to this.

This example shows that a high decentralized coverage is possible with appropriate design of the generation capacities from wind energy and PV. The power from wind energy and PV must be designed higher compared to the load to achieve a sufficient coverage rate. The role of photovoltaics in the coverage rate is often underestimated.

Table 7.5 shows an example of load coverage for different generation scenarios. The favorable generation variant with 75% of the annual energy from low wind and 25% from PV results in an annual high coverage rate of 81.7%. The cellular grid load consists equally of household loads and electromobility. The annual generation from wind and PV amounts to 130% of the total load. The storage capacities in MWh in the grid are related to the total generation capacity in MW and can store peak generation for one hour. This is a relatively small storage capacity that is already having a big impact.

If the PV share is replaced by wind energy, as shown in the middle column, the coverage rate drops to 69.2%. If there is no stationary storage capacity in the grid, that can also be used for wind energy, as shown in the last column, the coverage rate drops to 64.4%. From this, the combined annual generation characteristics of wind energy and PV are favorable for a high coverage rate. Even relatively small storage capacities also significantly improve the coverage rate in the distribution grid.

The storage units in the electric vehicles are charged according to the charging profile "at home and at work" and there is no provision for feedback of energy from the electric vehicle into the network. The electric vehicles therefore represent pure loads.

Figure 7.16 shows the annual duration curve of the difference between generation from wind energy and photovoltaics and the demand for residential units and electric vehicles.

Since the time series of the demand and the generation were compared with each other, these are the real durations of the over- or under-production. Accordingly, the duration of the overproduction period is 57% of the year and the underproduction 43%. If the ratio between generation and load were to be reduced, the entire curve would shift downwards.

The area with over-generation is larger than the area with under-generation. Due to the long-term periods without sufficient wind energy and PV, an expansion

Table 7.5 Variants of the annual coverage rate

Annual energy shares	Annual energy	Annual energy	Annual energy
Wind energy share	75%	100%	100%
PV energy share	25%	0%	0%
Generation/load (annual energy)	130%	130%	130%
Storage in hours of wind and PV power	1 h	1 h	0 h
Annual coverage rate	81.7%	69.2	64.4%

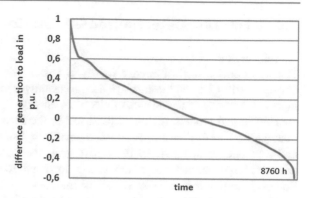

Fig. 7.16 Annual duration curve of the difference between generation and load

of storage capacities is not economical. Such capacities would only be used for a short time and would be uneconomical due to the high fixed costs [S4MG 2011].

In this study, only stationary accumulators were used, which can only take up small amounts of energy for a short period of time as daily storage and have low fixed costs per charging cycle due to the frequent charging and discharging cycles.

Instead of long-term storage capacities with low usage hours per year, power plants are more economical from today's standpoint of storage technology, even if they only have annual usage hours of 1000 to 2000 h/a. They can also close the generation gap over several months in the event of a long-term lull in wind generation, which only occurs once every 100 years, with a simultaneously low solar supply.

This is neither economically nor environmentally relevant through storage expansion. From today's perspective, power plant reserves in the order of magnitude of the highest grid load are therefore essential components of a secure energy supply.

In this example, the planning of the renewable decentralized energy supply was shown. In general, for analysis of the renewable potential time series with quarter-hourly supply values from wind energy, photovoltaics, hydropower, and biomass plants recorded over several years are required for the analysis of the renewable potentials. The total spatial potential for wind generation is to be determined from the approved areas. The spatial planning results in restrictions on the usable areas, e.g., through flight paths to airports, distance rules to inhabited buildings or prohibited areas in nature conservation areas.

PV installations on buildings may be regulated by local building codes and may not be possible, for example, in the case of listed buildings.

Since the renewable energy supply can only be economically implemented in efficient buildings with low heating energy demand and with efficient electric vehicles, the energetic renovation of settlements must go parallel to the planning of a predominantly sustainable energy supply.

7.8   Future of Decentralized Grid Services

Requirements

The traditional electrical energy supply is characterized by energy delivery with constant tariffs. For end user, there are no incentives to adjust his demand according to the scarcity or abundance of electrical energy.

Transitional models such as those used in the Smart Grid attempt to centrally manage end-users demand at constant tariffs to balance renewable supply with end-user energy demand. In the smart grid, the energy service provider often attempted in usual manner, to supply top-down and to bill the flow in the distribution grid through meters to the end consumer.

The future of electrical energy supply is likely to move more towards decentralized generation and storage at the end customers. This will require the development of new models, often associated with the Micro Grid. The following changes are to be expected in the future:

- Subsidies for renewable energy are being replaced by market models. If everything becomes renewable, not everything can be subsidized anymore.
- End user with photovoltaics will achieve a significant self-generation rate.
- Due to the reduction in the share of nuclear and fossil thermal power plants and a strong expansion of renewable energy sources, generation fluctuations will increase strongly.
- Since renewable energy sources with wind or photovoltaics have significantly lower full-load hours, higher generation power is required than with conventional power plants. The generation capacities can therefore at times be significantly higher than the network load. This can lead to overloading of the electricity grids, especially if they cannot be adequately expanded.
- For balancing the surplus energy, central storage capacities are required, especially as a central pumped storage power plants and decentralized accumulators in stationary or mobile applications.
- Renewable sources must participate more in grid services and be temporarily curtailed when end-use is not possible.
- The end customers must be able to adapt their consumption behavior more closely to the fluctuating generation situation and possible bottlenecks due to over- or under-generation. This requires flexible consumers who can adjust their consumption behavior more strongly according to a fluctuating load behavior, but who can make their own decisions and are not necessarily controlled by central facilities.
- Digitization and decentralization of energy supply with variable are ways to achieve these goals.

Digitization and variable tariffs

Digitization has centralized and decentralized forms.

The interconnected **smart meter** has a central functionality and establishes the connection between the end customer and the energy supplier or energy service provider. This enables the following subtasks:

- Counting the final energy requirement
- Counting the energy fed back
- Recording of load profiles
- Transmission of variable tariffs for purchase and feedback of the end customer
- Congestion management of the energy service providers through temporary load management at the end customer with corresponding counting of the remuneration

The automation system of the **smart home** represents the decentralized component of digitization. This receives tariff and congestion information from the smart meter. A decentralized load management can be derived from this based on economic incentives or to avert supply congestions and supply risks. The decentralized automation system can be freely selected by the end customer and is not subject to any central supervision. This increases data security and reduces supply risks through subversive activities. If possible, the smart home should communicate as end devices via radio-based local data networks to achieve greater flexibility for mobile end devices as well.

7.9 Summary

The development of decentralized supply structures, especially as energy cells with a high degree of self-generation from photovoltaics and decentralized wind energy, make it possible to relieve the central transmission grids and with the resulting reduced need for expansion, to operate them more economically. The decentralized supply structures as Micro Grids should be adapted in their decentralized generation capacities to the local demand to avoid high loads on the distribution grids when exporting surplus energy. Local stationary storage systems can be designed together with adapted dimensioning to cover about 50% of the local demand. Together with decentralized low-wind turbines, coverage rates of up to around 80% of the annual energy demand can be achieved economically. The characteristics of the final energy demand of households and electric vehicles should be flexible and more closely adapted to the variable decentralized energy sources.

A mixing ratio of the decentral generated energy should consist of around 75% of low wind systems and 25% of photovoltaic systems with storage. This results in the highest coverage rate over the year and reduces the need to expand the distribution grids. Furthermore, incentives for a more supply-oriented demand are possible by introducing smart market via market mechanisms.

Even with expansion of decentralized energy cells as Micro Grid, the central transmission networks with power plants are still necessary for grid control and providing balancing energy since renewable energy supply does not allow full supply due to its volatile character with wind lulls and dark periods.

By shifting PV generation to decentralized structures and by introducing low-wind systems inland, a strong, power-oriented expansion of the grids with uneconomically short usage periods can be avoided.

References

[ADRES 2009]   ADRES Concept: Autonome Dezentrale Regenerative Energiesysteme. Gefördertes Forschungsprojekt des Österreichischen Klima- und Energiefonds (2010)
[BDEW 2008]   Standardisierte Lastprofile (ehemals VDEW.Lastprofile)
[BNA 2011]   „Smart Grid" und „Smart Market" – Eckpunktpapier der Bundesnetzagentur zu den Aspekten des sich verändernden Energiesystems. Bundesnetzagentur Bonn (Dezember 2011).
[Brauner 2008]   Brauner, G., Leitinger, C.: Solare Mobilität 2030 – Machbarkeitsstudie zur solaren Vollversorgung im Verkehrsbereich 2030. Auftragsstudie des Lebensministeriums Wien (Mai 2008).
[DG DemoNet 2015]   DG DemoNetz – Smart LV Grid: Control concepts for active low voltage network operation with a high share of distributed energy resources. Forschungsprojekt des Klima- und Energiefonds (2015)
[Herry 2008]   Herry, M., Russ, M., Schuster, M., Tomschy, R.: Mobilität in Niederösterreich – Ergebnisse der landesweiten Mobilitätsbefragung 2008. Amt der Niederösterreichischen Landesregierung (2008)
[Leitinger 2011a]   Leitinger, C.: Netzintegration von solarer elektrische Mobilität – Auswirkungen auf das elektrische Energiesystem. Dissertation an der Technischen Universität Wien (2011)
[Leitinger 2011b]   Leitinger, C., Litzelbauer, M.: Netzintegration und Ladestrategien von Elektromobilität. e&i, Heft 1–2 (2011).
[SEM 2011]   Smart-Electric-Mobility – Speichereinsatz für regenerative Elektrische Mobilität und Netzstabilität. Forschungsprojekt des Klima- und Energiefonds (2011)
[S4MG 2011]   Super-4-Micro-Grid – Nachhaltige Energieversorgung im Klimawandel. Forschungsprojekt gefördert vom Österreichischen Klima- und Energiefonds (2011)
[Web2Energy 2012] EU Forschungsprojekt im Rahmen des EU-Rahmenprogramms FP 7. Projektleitung HSE Darmstadt (2012)

Efficiency of Buildings and Heating Demand

8

8.1 Residential Building Stock by Efficiency Classes

The buildings are subdivided into residential buildings and non-residential buildings. The residential buildings, in turn, consist of single-family and two-family houses (OTFH) and multi-family houses (MFH). Single-family and two-family houses are predominantly located in rural areas, in small towns and on the outskirts of large cities. The multi-family houses dominate in the city center.

German building stock
Table 8.1 shows the shares of the building types in the German stock and the share in of the energy demand (dena 2016).

The majority of the building stock, 71%, are single and two-family houses. The energy demand of the buildings for heating and air conditioning depends on the year of construction and the building type. Single-family buildings have a larger exterior envelope per residential unit (RU) compared to multi-family buildings. New buildings are required to comply with lower building energy limits, in accordance with the European directive 2010/31 *on the energy performance of buildings* [EU 2010/31], which are being transposed into national law. This has led to a significant reduction in the specific energy demand of new buildings [dena 2016], [IWU 2010].

The typical energy demand for space heating per m^2 of living space in kWh/m^2/a is shown by the year of construction in Table 8.2. Single-family houses and two-family houses have similar characteristic values. For older buildings, the values for multi-family houses are lower than for single and two-family houses. For new buildings, there is a convergence of values. The numerical values represent typical average values.

In 2016, the existing building stock had an average heating demand of 160 kWh/m^2/a. As a result of the EU thermal insulation directive, new buildings as low-energy houses have a specific heating demand of 50 kWh/m^2/a from the year

© Springer Fachmedien Wiesbaden GmbH, part of Springer Nature 2022
G. Brauner, *System Efficiency by Renewable Electricity*,
https://doi.org/10.1007/978-3-658-35138-0_8

Table 8.1 Building types and shares in the energy demand of buildings in Germany [dena 2016]

	One-/two-family house	Multi-family house	Non-residential buildings	Total of all buildings	Total residential buildings
	OTFH	MFH	NRB		
Millions	15.6	3.2	3.0	21.8	18.8
Shares building	71%	15%	14%	100%	86%
Energy shares	39%	24%	37%	100%	63%

Table 8.2 Final energy demand of residential buildings by year of construction [dena 2016; BMWiE 2014]

Building	Before 1919	1919–1948	1949–1978	1979–1990	1991–2000	2001–2008	2009–	Total
Share in the stock	13%	12%	38%	13%	14%	7%	3%	
Residential buildings in million	2.444	2.256	7.144	2.444	2.632	1.316	564	18.800
Heating energy demand mean values	Before 1919	1919–1948	1949–1978	1979–1990	1991–2000	2001–2008	2009–	2050
kWh/m²/a								
OTFH	250	260	230	175	130	80	50	20
MFH	200	210	170	140	120	70	50	20

of construction 2009. By 2050, this demand is expected to drop to passive house standards of 10 bis 20 kWh/m²/a or below.

> The rate of new construction is about 1% per year. By 2050, about 40% new residential buildings are expected to be built. By then, 50% of the existing building stock will probably also either have been replaced by new buildings or provided with thermal insulation to reduce the energy demand.

The average living space in new buildings is 140 m² for single-family houses and semi-detached houses and 80 m² for apartments in multi-family houses [dena 2016].

The heating systems installed in new homes between 2000 and 2015 are shown in Table 8.3.

Coal-fired heating has not been installed in new buildings since 2000. Oil-fired heating systems, which previously dominated heating systems, have also gradually

Table 8.3 Heating systems in new apartments [dena 2016]

Heating system	2000	2010	2015
Wood, pellet stove	0.0%	7.5%	6.5%
Heat pump	0.5%	29.0%	30.1%
District heating	4.1%	5.0%	7.1%
Electricity	1.5%	1.5%	1.1%
Solar heating	0.0%	1.0%	0.7%
Gas heating	74.3%	52.7%	52.4%
Oil heating	19.6%	3.3%	2.0%

been almost completely displaced. Heat pumps have gained significant shares since 2010. They are also displacing gas heaters. Heat pumps are used in particular in single-family houses. Here, there is often sufficient land area available for ground collectors or deep collectors, which are necessary for water-to-water heat pumps. Air–water heat pumps do not require geothermal collectors and can also be installed in or next to buildings.

Wood and pellet heating systems have a constant share. Due to the higher effort for feeding with fuel and disposing of ash, they are less suitable for long-term fully automatic operation.

Heat networks are possible as district or local heating networks in urban and rural regions. A minimum load density of the heat demand is necessary for an economic operation of the heating network.

Austrian building stock
The heating systems of one- and two-family houses in Austria by year of construction are shown in Fig. 8.1. Austria has a high proportion of wood, wood chips and Pellet (biomass) heating systems. Pellet stoves are very common and special research institutions are working on optimizing them. Heat pumps have been used since 1990 with an upward trend. Biomass heating and heat pumps account for two-thirds of all new single-family and two-family houses built from 2001

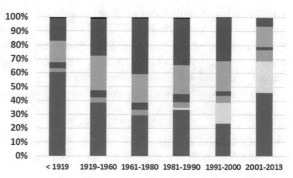

Fig. 8.1 Heating systems by year of construction for single and two-family houses [NEEAP-AT 2014]

Fig. 8.2 Heating systems
by number of apartments for
new buildings 2001–2013
(RU: Residential Unit)
[NEEAP-AT 2014]

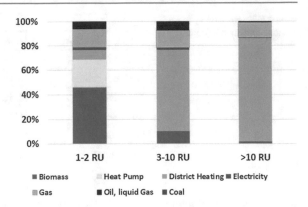

onwards. Coal-fired heating systems are no longer being installed. In some provinces of Austria, oil heating systems are no longer subsidized.

Figure 8.2 shows a breakdown according to the number of residential units
(RU) per building. According to this, heat pumps have so far not found any significant widespread use in apartment buildings. Biomass heating systems also have
a low penetration rate here, because they require caretakers for the operation in
case of central systems or apartments with chimney connection in case of individual stoves. The share of district heating and local heating systems is very high,
especially in apartment buildings. Heating networks have been heavily promoted
and expanded in large cities and also in smaller towns over the past few decades.
Cogeneration from thermal power plants is especially in the cities very common.
In rural regions, biomass local heating networks have been developed.

*Due to the widespread use of district heating and local heating connections, the medium-
term development potential for heat pumps in Austria is particularly in the area of single
and two-family houses.*

8.2   Energy Demand of Buildings

The required heating capacity of buildings is determined according to the
European standard DIN EN 12831. The heating power requirement is determined
by the building envelope A, the temperature difference between the inside temperature t_i and the outside temperature t_a and the average heat transfer coefficients of
the outside envelope shell U [Bauphysik 2007].

The transmission losses $\dot{Q}_H$ of the building envelope are

$$\dot{Q}_H = A \cdot U \cdot (t_i - t_a) = H \cdot (t_i - t_a) \tag{8.1}$$

The heat transfer coefficient U has the dimension W/K/m². The thermal conductivity value H in W/K determines the total heat flow through the building envelope

$$H = A \cdot U \tag{8.2}$$

Standard values apply to temperatures in accordance with DIN EN 12831. For temperate zones, the lower outside temperature is in the range of -12 to $-16\,°C$, in extreme situations such as the Zugspitze it is $-24\,°C$. The average outside temperature in the heating period is 4 to 7 $°C$. The Standard indoor temperature depends on how the room is used and is usually around 20 $°C$.

Additional heating is needed to compensate for the ventilation heat losses.

$$\dot{Q}_L = n_{min} \cdot V \cdot c_p \cdot \rho \cdot (t_i - t_a) \tag{8.3}$$

The air volume V of the living spaces, the specific heat capacity of the air at constant pressure cp, and the density of the air ρ are used together with the temperature difference. The exchanged air flows in from the outside and is heated to room temperature. The number of air changes n_{min} represents an empirical value, which is 0.5 to 3 complete air changes per hour.

Since the heat in buildings is reduced at night, the heating capacity must be dimensioned higher in order to return the building substance, which has cooled by a few degrees Celsius, to the standard room temperature in a given time of one to three hours.

In accordance with DIN EN 12831, the required maximum heating power for the lowest outside temperature specified in the standard can be determined. In case of electric heating, from the point of view of the utilities, in addition to the maximum heating power, which can occur simultaneously and is required when designing the distribution grid, the annual energy demand is also important. This not only depends on the technical design of the heating systems, but is also significantly influenced by user behavior.

The average temperature level of the apartment, the frequency of air changes, which are higher for smokers than for non-smokers, for example, and other usage habits of the residents are important. The specific annual energy demand in kWh/m²/a is required for the energy certificate of a building. This value can be calculated precisely from the annual consumption of energy sources used for heating, such as oil, gas, biomass or electricity, or from the heat demand, which is fed from a heating network. By the year 2050, the heating demand is to be converted largely to renewable energy in accordance with the goals of the energy transition. Since the potential of renewable energy are limited, these goals require significant efficiency improvements in the sector of building heating.

8.3 Heating Energy Demand

A variety of different heating systems are available. Oil and gas heating use fossil fuels. The spread of coal heating is decreasing because of the high pollutant emissions and because the operation requires staff for charging and ash disposal. As part of the energy turnaround, fossil heating systems are being replaced by efficient systems that use electricity generated from renewable sources in particular. Heat pumps are predominantly used for this purpose. In buildings with high thermal insulation, electrical direct heating or storage heating will also be energetically expedient and economical in the future due to the low energy demand. Pellet heating systems using biomass are also widespread as renewable heating systems.

Since 2015, coal-fired heating systems have no longer been used in new residential buildings in Austria. According to Statistics Austria, in 2015 they only had a share of 0.2% in all heating systems in buildings. Oil heating systems still have a share of 16% and gas condensing boilers a share of 24%. Heat pumps have increased and now have a share of 52% in new buildings and a share of 9% of all heating systems. District heating has the greatest penetration in existing buildings at 28%. Wood and pellet heating systems have a share of 17%.

From the point of view of energy efficiency, the energy demand of buildings can be significantly reduced by the use of heat pumps. A single-family house with 200 m^2 of living space and four people as residents, each of whom is entitled to a proportional living space of 50 m^2 serves as a reference. The specific living space of 50 m^2 per person is also intended to be used for the analysis of apartment buildings and large residential complexes.

The average heating energy demand of buildings in 2015 was around 150 kWh/ m^2/a. This results in a heating system for a single-family house with an annual heat demand of 30,000 kWh.

Table 8.4 shows the bandwidth of the investment costs for the heat demand using different heating technologies. For comparison, a typical service life of 20 years is assumed for all types. For the economic evaluation of the systems, a contracting model is to be considered over the service life of 20 years with an interest rate of 3% on the capital employed. The annual maintenance costs are uniformly taken into account at 2.5% of the investment costs.

The energy costs for gas are assumed to be at a minimum of 7.5 €ct/kWh and at a maximum of 10 €ct. For heat pumps and direct electrical heating, the minimum

Table 8.4 specific investment costs for heating systems

Plant type 30,000 kW/a	Abbreviation	Investment costs	
		Min	Max
Gas heating/oil heating	GH	7500	10,000
Gas solar heating	GSH	17,500	24,000
Brine-water heat pump with geothermal stakes	BWGS	14,850	19,850
Brine-water heat pump with geothermal collector	BWGC	11,500	15,000
Water-to-water heat pump	WW	17,000	19,000
Air-to-water heat pump	AW	13,000	16,000
Air-to-air heat pump	AA	7000	12,000
Electric underfloor heating (<100 €/m^2)	EF	14,000	18,000
Electric storage heater	ES	5000	8000
Electric direct heating	ED	3000	5000

energy costs are assumed to be 24 €ct/kWh and the maximum to 40 €ct/kWh. This seems realistic, since the heating systems are mostly in operation in winter. During this period the available solar radiation is very limited. In contrast, there is good wind supply, which is transmitted to the end customers via the grid levels and is therefore subject to grid tariffs.

8.4 Heating Costs and Electricity Demand

8.4.1 Monovalent and Bivalent Gas Heating with Solar Thermal Energy

Figures 8.3 and 8.4 show the annual cost from amortization rate, energy price and maintenance costs for **monovalent gas heating systems** and **bivalent gas heaters with solar thermal energy**. The solar thermal energy reduces the fossil heating energy demand by around 20% in winter. Bivalent systems are not very economical because of the high installation costs.

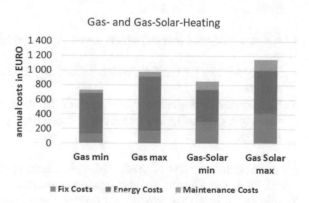

Fig. 8.3 Energy costs for gas heating systems in residential unit of 150 kWh/m²/a

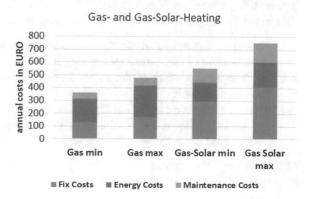

Fig. 8.4 Energy costs for gas heating systems in residential unit of 50 kWh/m²/a

Figure 8.3 show the total heating costs for an average of 150 kWh/m^2/a and a living space of 50 m^2.

For comparison, Fig. 8.4 shows the annual costs for the same living space if the building is thermal insulated to reduce the heating demand to 50 kWh/m^2/a. Due to the high investment costs of gas heating systems with solar thermal supplement, the savings in fossil are compensated by higher depreciation of bivalent systems for the latter type of system.

8.4.2 Heat Pumps: Theory and Designs

Heat pumps make use of environmental heat from the air or ground for heating buildings or domestic hot water by "pumping" the energy latently present in the environment to a higher temperature level that meets the technical requirements. For example, the heat from the outside air is raised from 10 °C to a level of 35 °C corresponding to the flow temperature of an underfloor heating.

Heat pumps are differentiated according to the heat carriers from which the energy is extracted. The heat is usually fed into a water circuit to heat the building. A distinction is made between

- Air-to-water heat pumps, which extract the thermal energy from the ambient air,
- Water-to-water heat pumps, which extract thermal energy from the groundwater and
- Brine-water heat pumps, which extract the thermal energy from the ground via geothermal collectors or probes and use frost-proof brine as a heat transfer medium.

From the point of view of thermodynamics, heat pumps represent a left-running cycle. First, a clockwise circular process is considered for explanation (Fig. 8.5, T,s-diagram). In a clockwise steam turbine cycle, heat is supplied and mechanical work is output. An example of this is the clockwise Clausius-Rankine process [Baehr 1989, Cerbe 2005] as a comparison process for steam power plants. It consists in the T,s-diagram of the following sub-processes (Fig. 8.5):

3-2   isobaric heat supply in the steam generator for preheating, evaporation and overheating of the water as a working fluid

2-1   isentropic expansion of the steam in the steam turbine with the output of mechanical energy

1-4   Isobaric heat removal to condense the steam in the condenser (in the steam cycle to the liquid line to the left), then compression of the water in the boiler feed pump to the boiler pressure 3.

The degree of effectiveness of the process is described by the mechanical energy released in relation to the heat absorbed.

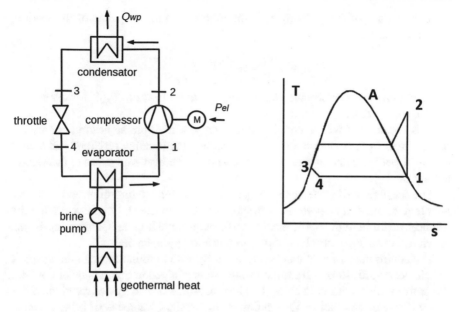

Fig. 8.5   Thermal schematic and T,s-diagram of the heat pump process

According to Fig. 8.5, the counterclockwise heat pump process consists of a compressor, a condenser to deliver the heating heat, a throttle valve, and an evaporator where environmental heat is absorbed. The thermodynamic process consists or the following steps:

1-2   Reversible compression in the adiabatic compressor. This increases the pressure and temperature of the refrigerant.
2-3   First, isobaric cooling, then along the horizontal line isobaric condensation and then further cooling of the liquid refrigerant to point 3. Curve A represents the phase boundary conditions of the working fluid. The left branch is the boundary line to the liquid state area. In this area, heat is dissipated. The right sloping branch represents the boundary line to the superheated steam. The area between the left and right branches represents the wet steam area with partial condensation.
3-4   Adiabatic throttling with partial evaporation of the refrigerant.
4-1   Complete evaporation of the refrigerant up to state point 1. Ambient heat is absorbed in this area.

The coefficient of performance ε_{wp} of the heat pump represents the ratio of the heat delivered Q_{WP} to the electrical energy consumed to drive the compressor W_{el}. Frequently, the heat $\dot{Q}_{WP}$ output as heat energy per time unit and the electrical drive power P_{el} are also used. The coefficient of performance can be calculated from the values of the enthalpy h_i in the individual process points of the working medium (Fig. 8.5).

$$\varepsilon_{wp} = \frac{Q_{wp}}{W_{el}} = \frac{\dot{Q}_{wp}}{P_{el}} = \frac{h_2 - h_3}{h_2 - h_1} \tag{8.4}$$

The coefficient of performance has a value greater than one. Typical working values are from 3 to 6.

This means that with a certain drive energy, a multiple as heating energy can be obtained. In comparison, direct electrical heating (heating resistors) has a coefficient of performance of one, i.e., this produces the heat equivalent to the electric energy used.

The coefficient of performance depends on the temperature difference from the heat transfer medium between the evaporator and condenser. Low temperature differences result in high coefficients of performance and high temperature difference in winter cause the coefficient of performance to approach one.

The environmental heat can be absorbed from the ground through the evaporator via deep earth stakes, from the groundwater via geothermal collectors or from the ambient air via air collectors. The heat obtained is usually delivered via radiators with water circulation. Depending on the design, we speak of brine-to-water, water-to-water or air-to-water heat pumps.

Table 8.5 shows the coefficient of performance for water-to-water heat pump as a function of the temperature difference between the groundwater and the flow temperature of the heating system.

Small temperature differences mean high performance figures. Over the year, water-to-water heat pumps have a higher average coefficient of performance, because the groundwater is in a relatively constant temperature range of 10 to 15 °C. For air-to-water heat pumps, the air temperature in regions with temperate climates can range from −25 °C to +35 °C. The annual performance figures are therefore lower.

Air-to-water heat pumps are easy to install on the roof, in the basement of buildings or in open spaces, and are easier to obtain permits for. Therefore, they have a wide distribution. Water-to-water heat pumps with collectors in the groundwater or brine-to-water heat pumps with ground stakes often require extensive official approval procedures and have fewer potential for application in urban areas. There is a higher potential for use in rural areas.

As Fig. 8.6 shows, high coefficients of performance result at low flow temperatures of the heating system and at high ambient temperatures of the heat collector. For an economical design, the overall system consisting of thermal insulation

Table 8.5 Coefficients of performance of water-to-water heat pumps as a function of temperature difference

Temperature difference	60	50	40	30	20	10
Coefficient of performance	2.8	3.2	3.9	5.1	7.4	10.0

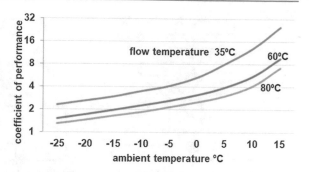

Fig. 8.6 Coefficient of performance of a heat pump depending on the ambient and flow temperature

measures on the building, design of the heating surfaces of the radiators and heat output of the heat pump must be optimized for the planned minimum outside temperature.

At very low outside temperatures, which only occur a few days a year, additional heating is often more economical than oversizing the heat pump.

For economical and effective operation, the following must be observed when using heat pumps:

- A low difference of the supply temperature to the external ambient temperature can be achieved by low-temperature floor heating systems in buildings with thermal insulation.
- Increase in the effective heating surfaces of the radiators can also enable the same heating output at a lower flow temperature.
- The selection of sunny locations of the evaporators for solar preheating of the air in case of air-to-water heat pumps or of the ground area in which the heat exchangers are located in case of water-to-water heat pumps, is advantageous.
- At low ambient temperatures, the coefficient of performance approaches one, i.e., the heat pump then acts only as a direct electric heater.

Figure 8.7 shows the principle of a brine-to-water heat pump with a geothermal collector. In the ground collector, a water–glycol mixture is used as a heat transfer medium to ensure frost-proof operation. Underfloor heating systems with a low flow temperature of maximum of 35 °C are used in the building. The large available heating surface of the living space allows low flow temperatures. As a result, the heat pump has to overcome a lower temperature difference and a correspondingly higher coefficient of performance. Air-to-water heat pumps can be integrated in the building with supply air and exhaust air ducts and can be used in urban residential areas. In general, geothermal collectors cannot be installed on public land.

Average annual performance figures are used for planning of heating systems. These depend on the mean outside temperature, the flow temperature of the heating system and the average temperature of the heat collector. With brine-to-water heat pumps and water-to-water heat pumps, the annual coefficients of performance

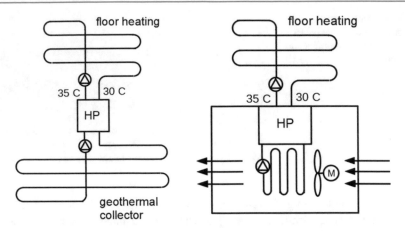

Fig. 8.7   Brine-to-water heat pump with ground collector and air-to-water heat pump

are higher than with air-to-water heat pumps, since the ground has a higher temperature level than the ambient air over the year.

The annual electric energy demand of a heat pump $W_{el,a}$ can be calculated with the annually generated amount of heat $Q_{wp,a}$ and the average annual coefficient of performance $\varepsilon_{wp,a}$.

$$W_{el,a} = \frac{Q_{wp,a}}{\varepsilon_{wp,a}} \tag{8.5}$$

When planning of heat pump systems, lower values are usually used for the annual coefficients of performance in order to have reserves. Typical lower values for annual performance figures are $\varepsilon_{wp,a} = 3.5$ for air-to-water heat pumps and 4.0 for brine-to-water heat pumps and 4.5 for water-to-water heat pumps [Bonin 2015].

8.4.3   Design of Heat Pumps for Space Heating

It is assumed that buildings constructed before 1990, have undergone thermal insulation measures, e.g., the replacement of windows were carried out and thus the specific heating demand does not exceed 150 kWh/m²/a.

Table 8.6 shows the average specific heating energy demand in kWh/m²/a and the annual average heating energy demand for a single-family house or row house with 150 m² of living space and the planning values for the heating capacity per m² of living space.

First, the heating energy demand for medium-sized buildings with a specific heating energy requirement (W) of 150 kWh/m²/a is to be calculated. The heat pump used should be designed according to Fig. 8.6 for a forward flow temperature (FFT) of 60 °C and for hot water heating. The air-to-water heat pump has a heating output of 20 kW at an ambient temperature of 15 °C (Ref AT).

Table 8.6 Heating demand and heating capacity by building class

Year of construction or heat insulation regulation	Specific heating energy demand	Specific heating power at −5 °C	Heating power at ambient temp. 15 °C space 160 m²	Annual heating energy demand $Q_{wp,a}$
	kWh/m²/a	W/m²	kW	kWh
Passive house	10–20	5–10	5.0	1600
2009–2011	50	25–50	20.0	8000
Medium stock	150	70–120	40.0	24,000

Figure 8.8 shows how the heating power of the heat pump increases with the outside temperature. In this example, the room temperature (RT) is 20 °C. The heating demand of the house decreases proportionally to the ambient. At −5 °C, the two curves intersect. This means that below this ambient temperature the heating output is not sufficient for this design.

In real operation, the outside temperature is only briefly below this value during night frosts and recovers during the day. In addition, the masonry represents a heat store with a cooling time constant of around 20 to 70 h. During the day at higher ambient temperatures, the heat pump can compensate for the deficit. However, during longer frost periods, the heating capacity is not insufficient and additional measures may have to be taken:

- Use of electrical auxiliary heaters. This is economical if only a few frost days occur in regions with a moderate climate.
- Temporary lowering the room temperature, e.g., to 18 °C. The is at the expense of comfort.
- Increase the heat pump capacity. At 30 kW, for example, the point of intersection shifts to -15 °C. This is necessary in regions with a low average temperature level, for example in Alpine valleys or high altitudes in mountains.
- This may also be necessary in windy regions, since the building envelope has greater heat losses.

Fig. 8.8 Heating demand and heat pump capacity

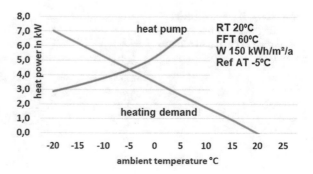

Fig. 8.9 Annual duration curves of the temperature in the high mountains and in central European regions

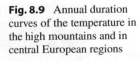
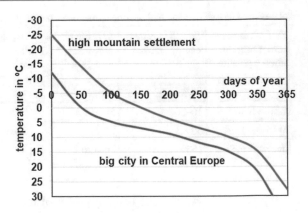

Figure 8.9 shows the annual duration curves of ambient temperature for two regions, one represents a high mountainous region, the second applies to Central Europe climates.

Temperatures below −5 °C can occur in large Central European cities in about 25 days per year [IWU 1996]. Since here is an ordered annual duration curve of the ambient temperature, the periods of low temperature are actually much shorter and are interrupted by periods of warming.

In high mountain settlements over 1000 m above sea level, 100 days with temperatures below −5 °C are possible. Long periods at low temperatures can occur. A higher output of heat pumps is advisable here.

Up to now, the use of heat pumps in the existing buildings stock with average characteristic values has been investigated. Now, the savings potentials through more energy-efficient construction are to be examined. For this purpose, the heat demand of a detached single-family house according to passive house standard with 20 kWh/m²/a is considered (Fig. 8.10).

The flow temperature can be lowered to 35 °C here by using a new underfloor heating system. For comparison, two air-to-water heat pumps with heating capacities of 4 kW and 2 kW at 10 °C are shown. The heat pump of higher power manages without additional heating all year round. In buildings with a low heat

Fig. 8.10 Heating demand of a single-family house according to the passive house standard

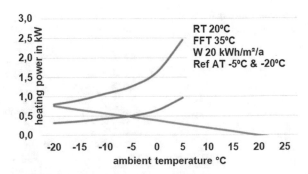

demand, a slight oversizing the heat pump is recommended, as the energy demand is low.

8.5 Heat Pumps for Room Cooling

Heat pumps can be modified so that they can also be used for room cooling in summer. For this purpose, the radiators are used to absorb heat and the heat collectors, earth stakes or air heat exchangers give off heat to the environment. The coefficient of performance as a refrigerator is according to the left-turning cycle in T,s-diagram in Fig. 8.5, shown in enthalpy h of the working fluid

$$\varepsilon_{wp,cool} = \frac{Q_{cool}}{W_{el}} = \frac{\dot{Q}_{cool}}{P_{el}} = \frac{h_1 - h_4}{h_2 - h_1} \tag{8.6}$$

In order to be able to use a heat pump system for heating and cooling, it must be equipped with a four-way valve. Figure 8.11 shows the application for heating on the left and cooling on the right.

By reversing the four-way valve, the direction of flow in the compressor is maintained, whereas in the throttle valve it is reversed. The condenser and evaporator now swap their functions. In summer, the system can also be operated with electricity from photovoltaics and thus represents solar cooling.

8.6 Direct and Storage Heating

Storage heaters were widely used as night storage heaters in the period from 1960 to 1980. They were charged at night with surplus energy from large coal-fired and nuclear power plants, which enabled them to drive through in a more

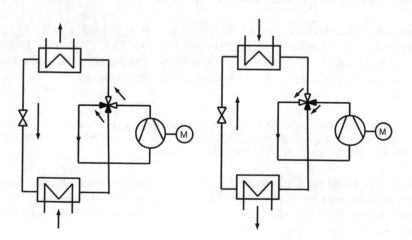

Fig. 8.11 Heat pump for heating and cooling

efficient operating state. As electricity tariffs have risen in the meantime, storage heaters are no longer economically viable in buildings with low thermal insulation standards.

In the future, conditions will change again in favor of electrical storage heating. The expansion of photovoltaics and especially wind energy will lead to periods of high surplus generation in Europe from 2020 onwards. The coupling of the sectors electricity and heat creates an opportunity to use this surplus energy flexibly and to save electrical storage capacities. Electrical direct and storage heating systems can only be used economically if the heat demand of buildings is correspondingly low. As the renovation of existing building stock progresses, these heating systems can become more widespread.

Three types of electrical heaters can be distinguished:

- Direct heating. Here the heating power is provided directly. The resistance heaters are very inexpensive.
- Flat storage heaters. These heaters look similar to flat radiators in direct heating. They have a core of narrow storage stones, in which there are channels with heating resistance wires. The charging temperature is limited to about 100 °C. The charging or discharging time constant is around 15 to 30 min. They represent short-term storage with low stored energy in the range of 1 to 3 kWh.
- Storage heaters (former name: night storage heaters) have a nominal power of 1 to 8 kW and charging times of 8 to 10 h. The stored thermal energy then corresponds to 8 to 60 kWh. The storage heaters contain shaped stones in a thermally insulated housing, which can be heated up to 650 °C. A thermostat-controlled fan allows the heat to be dosed for room heating. Compared to the previous night storage heaters, the thermal insulation of the devices must be significantly improved in order to avoid unwanted overheating of living spaces in low-energy buildings due to radiant heat loss.

In the future, the third category of electric heaters, storage heaters, will be of particular interest with regard to renewable energy systems. In the European transmission grid, the significant expansion of wind energy without adequate expansion of the transmission grids is increasing bottleneck situations that lead to the shutdown of wind turbines.

By providing flexible loads with storage capacities that can be activated according to the supply, regional demand side management in the distribution grid with storage heating can relieve the transmission grid of long-distance transits.

In low-energy and passive houses with low space heating demand, storage heaters are well suited for load management. They are also economical because of the low annual energy demand. This coupling of the electricity and electricity heat sectors also saves electrical storage capacity in the grids.

Fig. 8.12 Heat demand and discharge duration for a low-energy house with 50 kWh/m²/a and a storage heater of 48 kWh with a nominal power of 6 kW

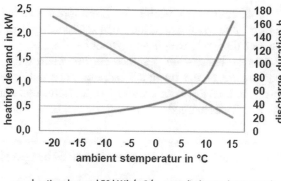

heating demand 50 kWh/m2/a ——discharge duration in h

In the future, tariff models are conceivable in which surplus energy can be purchased at low prices. If necessary, the storage heaters should be designed more strongly in terms of their storage time constants and nominal power in order to allow interruptions in recharging of several hours.

The classic design as night storage heating with a charging time of 8 h and a specific heating energy requirement of 50 kWh/m²/a at 50 m² is shown in Fig. 8.12. Even at a lower outside temperature of −20 °C the required room temperature can be maintained. For this purpose, the delivered heating power must be a factor of 2.4 (planning value) above the value from living space and specific heating demand per area element.

The core of storage heaters consists of shaped bricks with a high magnesite content, which are heated up to 650 °C (Fig. 8.13). The volume of the brick determines the storage capacity. For 45 kWh, storage bricks with a weight of 280 kg are required so that a storage heater including thermal insulation and control electronics is in the range of around 310 kg. Such performance is required for large rooms of around 150 m².

Fig. 8.13 Storage capacity as a function of the core temperature

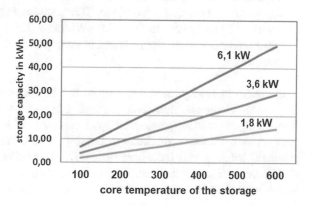

In the living area, each room is equipped with a storage heater, which, depending on the thermal insulation, is in the range of 1 to 2 kW with a charging time of 8 h, and then only weighs up to about 100 kg.

Storage heaters are designed so that when fully loaded to 600 °C, the surface temperature of the insulated radiator does not exceed about 60 °C. This design results in heat losses of around 1% per hour, i.e., after around 100 h the heat storage tank is discharged.

8.7   Future Requirements for Storage Heating

Night storage heaters were designed as day storage, which were charged at night and discharged during the day. This corresponded to the production rhythm of the thermal power plants. The apartments had lower thermal insulation standards and thus a higher heating demand. The thermal insulation of the storage heating systems was not subject to very high requirements, since the heat loss they gave off during the heating period was almost always lower than the required heating demand of the building. In order to limit the heating power according to the demand, they had a control device, that used a sensor for the outside temperature, to limit the charging temperature and thus the thermal load accordingly. The control system is set so that between two loadings the stored heat is sufficient to the supply heat, including reserves for reheating. This has the effect that lower storage capacities are available for demand side management at high outside temperatures.

In the future, the framework conditions for storage heaters will change:

- Renewably generated electrical energy from wind and photovoltaics has volatile generation characteristics. Longer periods of high generation may be followed by longer periods of little or no generation. Heat storage systems with capacities for several heating days up to a week can therefore be of interest with regard to building heating and the sector coupling of electricity and heat.
- In future, the buildings will have a very low specific energy demand due to the thermal insulation regulations. This will decrease from the current level of around 150 kWh/m²/a to 50 or 20 kWh/m²/a. The heat dissipation to the room must then not exceed the required heating power of the room in order to avoid overheating.

The system of storage heater and room can be represented in its dynamic behavior by an electrical equivalent circuit. Figure 8.14 shows these relationships. The storage heater is represented by a capacitor with the stored energy Q_S. The transmission losses of the storage heating system are represented by a thermal conductivity H_{ST} with the dimension W/K. The building envelope has a thermal conductance H_{GT} and the building itself a heat capacity of Q_G. This represents the stand-by mode of the storage system. The storage heater is charged with a constant heat power Q_L until an internal temperature level t_S specified by the controller is reached. In heating mode, a variable resistor H_{SH} is switched on, representing the

Fig. 8.14 Equivalent circuit
diagram of a heating system
with storage heating and
building

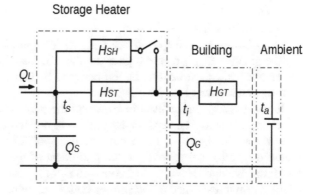

fan, which determines the heating output and regulates it to a constant room temperature t_i. The external environment of the building is represented by a source with a constant external temperature t_a, since even large amounts of heat given off there do not lead to a significant increase in the ambient temperature.

The heat capacity Q_G of the building together with the thermal conductivity value of transmission losses H_{GT} determines the thermal time constant T_G of cooling down of the building when the heating is switched off.

$$T_G = \frac{Q_G}{H_{GT} \cdot (t_i - t_a)} \tag{8.7}$$

For low-energy buildings, values in the range of 60 to 80 h are achieved. For reheating, a higher power is required than in stationary continuous operation.

In stand-by mode, the dissipation losses of the storage heater via the thermal conductivity H_{ST} of the thermal insulation must not exceed the heating demand of the building at the standard indoor temperature of $t_i = 20$ °C. For low-energy buildings and outside temperatures t_a of, for example, +5 or +10 °C with low heating demand, the indoor temperature t_S of the storage heater must therefore be set lower. However, this limits the possible stored energy.

In the case of classic night storage heaters, this was irrelevant, as the charging temperature was based on the demand of the next day and, with higher outside temperatures, there was also less heating energy required during the day. In the future, the aim is to store as much energy as possible. The permissible upper loading temperature is determined by the thermal conductivity value H_{ST} of the thermal insulation of the storage heater and the thermal conductivity value H_{GT} of the building envelope.

$$\dot{Q}_{ST} = H_{ST}(t_s - t_i) \leq H_{GT}(t_i - t_a) \tag{8.8}$$

This results in the permissible charging temperature depending on the outside temperature

$$t_S \leq \left(\frac{H_{GT}}{H_{ST}} + 1 \right) \cdot t_i - \frac{H_{GT}}{H_{ST}} \cdot t_a \tag{8.9}$$

Example:

$H_{ST}=0.96\ W/K\ H_{GT}=20\ W/K$

Outside temperature $t_a=-5\ °C: t_S\leq540\ °C; t_a=+5\ °C: t_S\leq332\ °C.$

The smaller H_{GT}, i.e., the higher the thermal insulation standard of a building, the smaller H_{ST} must be, that is, the better to be thermal insulation of the storage heating system must be.

Equation (8.9) can be used to construct the permissible internal temperature of the storage heater. The rays for two different outside temperatures t_a in Fig. 8.15 intersect at the specified room temperature t_i of 20 °C. The horizontal distance between t_i and t_a corresponds to the reciprocal of the thermal conductivity of the building envelope. The distance from t_i to t_{S1} or t_{S2} corresponding to the reciprocal of the thermal resistance of the storage tank insulation.

Figure 8.15 shows the temperature levels t_S in the storage heaters with differently thermal insulation. One storage heater is with normal thermal insulation t_{S1} as previously used, and the second with super insulation t_{S2}, that is, higher heat resistance $1/H_{ST}$.

At a relatively high ambient temperature of 15 °C, the heat loss of the storage heater in this example is sufficient to heat a building with a very high thermal insulation standard. The storage heater 1 (axis t_{S1}) may only be charged to 200 °C in order to avoid excessively high room temperatures, but the second with super insulation (t_{S2}) can be charged to 300 °C. This means that the better insulated storage heater can absorb larger amounts of heat and thus heat for longer period.

Storage heater 2 can be fully charged at the standard outdoor temperature of +5 °C without heating the room inadmissibly, storage heater 1 only to 57%. The relationships shown here in principle, should be taken into account when designing new storage heaters for low-energy houses.

Fig. 8.15 Temperature levels in the storage heater and in the building

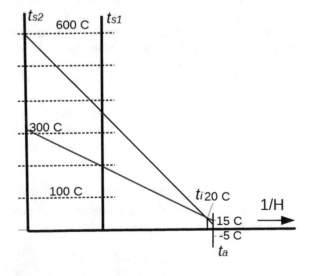

Table 8.7 Heat storage in comparison

Material	c_W	ρ	t_{max}	Specific volume	Storage volume at 50 kWh	Total weight at 50 kWh
	kJ/kg/K	kg/dm³	°C	kWh/dm³	dm³	kg
Magnesite bricks	1.0	3.0	600	0.483	104	350
Water	4.212	1.0	95	0.088	570	600

By improving thermal insulation (super insulation) of the storage heater higher internal storage temperatures and thus higher loadings are possible, as shown by the left axis in Fig. 8.15. Thus, storage heaters with capacities sufficient for a heating day at today's thermal insulation standards of buildings of 150 kWh/m²/a can be used as weekly storage in low energy houses with a standard of, e.g., 20 kWh/m²/a with a thermal super insulation. Such storage heaters therefore also offer capacities for demand-side management to relieve the network.

The heat storage capacity Q of a storage heater is determined by the mass m of the storage unit and the specific heat capacity c_W.

$$Q = m \cdot c_W \cdot \Delta T = \rho \cdot V \cdot c_W \cdot \Delta T \qquad (8.10)$$

A comparison of a solid-state storage of magnesite bricks with a water buffer storage tank is shown in Table 8.7. The total weight includes insulation.

At higher storage temperatures, storage systems with smaller volumes can be produced. Magnesite is stable up to 3000 °C. For safety reasons, and because of the limited temperature levels of the electrical heating elements, lower temperature levels down to 650 °C are chosen. Storage heaters can be placed as smaller decentralized units in every living room. Buffer tanks are suitable for central heating and hot water supply.

8.8   Efficiency Potentials and Cost-Effectiveness of Heating Systems

8.8.1   Annual Costs of Heat Pump Types in Comparison

The annual costs for energy are again determined per person, each of which is assigned a living space of 50 m². The service life of the heat pumps should be 20 years. Over this period, the invested capital should be repaid at an interest rate of 3% in equal annual installments over the service life. This represents the proportion of fixed costs. The minimum and maximum fixed costs result from the variation range of the investments according to Table 8.4. The annual maintenance costs amount to 2.5% of the investment values.

For energy costs, the minimum value e for electricity prices of 24 €ct/kWh and the maximum for higher electricity tariffs to be expected in the future of 40 €ct/

kWh are assumed. The annual performance factors of the heat pumps used are uniformly $\varepsilon_{wp,a} = 4.0$ for brine-water heat pumps with ground stakes (BWGS) and with ground collectors (BWGC). For water-to-water (WW) $\varepsilon_{wp,a} = 4.5$ and air-to-water (AW) and for air-to-air (AA) $\varepsilon_{wp,a} = 3.5$ applies.

Figure 8.16 shows the annual costs per person for 50 m^2 of living space and a heating demand of 150 kWh/m^2/a.

The annual electricity demand is compared later in Fig. 8.19 for various heating systems with heat pumps or direct electric heaters.

At a reduced heating requirement of 50 kWh/m^2/a there is again a significant reduction in the annual costs (Fig. 8.17).

8.8.2   Electrical Direct and Storage Heating

Direct electrical heaters are resistance heaters in which the electrical energy is directly converted into heat energy. The conversion efficiency is 100%. Compared

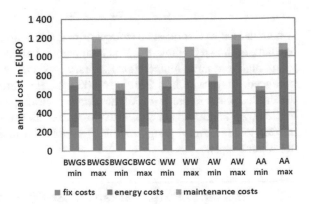

Fig. 8.16 Annual costs of heat pumps at 150 kWh/m^2/a

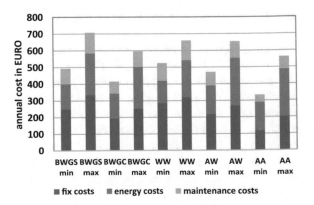

Fig. 8.17 Annual costs of heat pumps at 50 kWh/m^2/a

to heat pumps, the full heating energy must be provided by electricity. Heat pumps have a lower energy demand by a factor of the annual coefficient of performance and are therefore more efficient. Electric direct heaters require very little maintenance compared to heat pumps, gas heaters and solar thermal supplementary systems.

Storage heaters can temporally decouple the electrical energy demand and the heat demand. In direct heating, electricity and heat demand are temporal coupled. Storage heaters are much more adaptable to volatile regenerative generation.

Because of the low investment costs, direct heating systems are particularly advantageous in buildings with low heat demand. In contrast, electrical direct heating systems are not very economical in buildings with a low thermal efficiency standard. Figure 8.18 shows the annual costs for a living area of 50 m^2 and a heating demand of 150, 50 and 20 kWh/m^2/a, respectively, at electricity prices in 2015 of 0.24 €/kWh and in the year 2050 of 0.40 €/kWh.

In the case of electric storage heaters, the annual costs for electricity remain roughly the same, but the fixed costs double. Since the annual fixed cost components are low due to the long service life of the equipment of 20 years, the annual costs of storage heaters are not significantly higher than with direct heaters.

8.8.3   Electricity Demand of Heat Pumps and Direct Heating

In buildings with poor thermal insulation, direct heating and storage heating are uneconomical compared to heat pumps. The fixed costs from depreciation are low compared to the high energy costs for electricity (0.24 €/kWh). In the future, electricity prices will continue to rise, on the one hand due to investments in renewable generation systems, central energy storage and on the other hand due to the necessary expansion of the transmission and distribution grids.

Fig. 8.18 Annual costs of direct electric heating for 50 m^2 living space

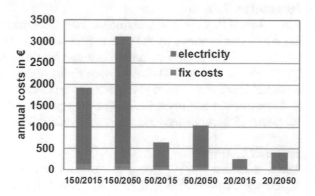

Fig. 8.19 Electricity demand
for direct heating and heat
pumps

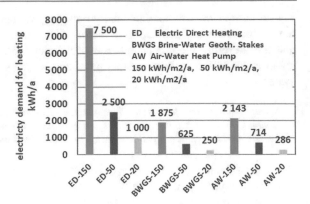

With good thermal insulation of buildings, energy costs remain very low, even with higher electricity prices. Therefore, electrical direct or storage heaters are economical here. From the perspective of the renewable electrical system, electric heaters are also useful in addition to heat pumps in order to be able to limit the expansion of expensive electric storage facilities such as pumped storage and accumulators in the case of renewable overgeneration through sector coupling to heat.

Figure 8.19 shows for direct electric heaters or storage heaters (ED), as well as for heat pump types for brine-water with ground stakes (BWGS) and air–water (AW) the annual electricity demand. This is again presented on a living area of 50 m^2 per person and various thermal insulation standards of 150, 50 and 20 kWh/ m^2/a. With a specific heating energy demand of 20 kWh/m^2/a and less, electric direct or storage heaters can be used economically in the passive house or zero-energy house because of the very low demand for electricity.

8.9   Efficiency Potential for Space Heating by 2050

Potential in Germany
It is assumed that the new construction rate accounts for 1% of the existing building stock. Furthermore, by the year 2050, the existing buildings should be reduced by thermal refurbishment from an average heating demand of 150 kWh/m^2/a to 50 kWh/m^2/a. Figure 8.20 shows a scenario in which a high share of heat pumps is assumed for new construction projects and renovated old buildings.

Under these conditions, the heating energy demand in Germany can be reduced to around 20% compared to the reference year 2000.

Potential in Austria
The demand for heating energy in Austria was presented in the National Energy Efficiency Action Plan [NEEAP-AT 2014] including a building renovation

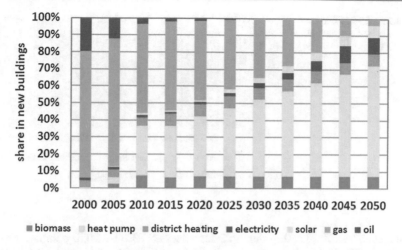

■ biomass   heat pump ■ district heating ■ electricity   solar   gas ■ oil

Fig. 8.20 Estimation of share of heating systems in new buildings in Germany

strategy. Subsequent studies assessing the potential for the use of high-efficient CHP and efficient district heating and cooling [TUW Ecofys 2015], [TUW 2018] present detailed potentials of future heating energy demand and energy carriers used. The values of this study were modified for the following scenario:

- In the building heating sector, coal will be phased out completely by 2025, oil heating by 2040 and gas heating by 2050.
- An annual thermal refurbishment rate or new construction activity totaling 1% per year is assumed. From the year 2020, the average thermal insulation standard is 50 kWh/m²/a. After that, it will decrease linearly by 2040 to the low building standard of 20 kWh/m²/a and will remain at this value until 2050.

Fig. 8.21 Scenario for heating energy demand in Austria until 2050

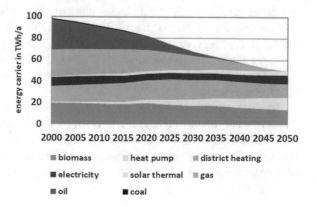

■ biomass   heat pump   ■ district heating

■ electricity   solar thermal   ■ gas

■ oil   ■ coal

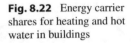

Fig. 8.22 Energy carrier shares for heating and hot water in buildings

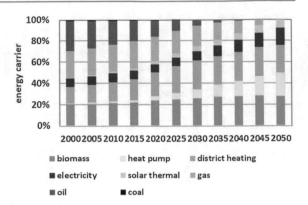

In this scenario, the heating demand decreases to 50% by 2050 compared to the year 2000 (Fig. 8.21). The Austrian scenario was and is characterized by a high share of biomass heating systems and a district heating. Traditionally in Austria heating systems based on logs, wood chips and pellets are very common.

District heating networks and local heating networks were promoted in accordance with the Austrian energy strategy. They will continue to expand further until 2020, and a slight decline is not expected until 2030 [TUW Ecofys 2015].

District heating network are mainly located in large cities and are mainly fed from existing new and efficient combined cycle gas and steam power plants, operated as combined heat and power (CHP). The few remaining coal-fired cogeneration power plants have already reached the lifetime limit and have been shut down since the year 2020.

Two-third of the energy sources used to generate district heating are fossil fuels and one third renewable energy.

Local heating networks are often operated with biomass and are in the range of a few megawatts. They are widely represented in Austria.

In the multi-family houses, the heat supply from heating networks and gas heating dominates. In newly built single- and two-family houses biomass, heat pumps and gas heaters are very common. This is likely to have the greatest growth potential for heat pumps, storage heaters and solar thermal systems.

In the future, electric storage heaters can also be used for flexible demand-side management in all types of buildings that have been refurbished with regard to thermal standards.

For the scenario, shown in Fig. 8.22, the assumed renovation rates of buildings result in a very low electricity demand for heating of only 20 TWh/a.

This is with a view to sustainable development of Austria extremely favorable, since the use of biomass for individual heating systems and heating networks conserves the potential of renewable electricity generation for use electromobility, industry and commercial trade and services.

8.10 Summary

The construction of low-energy houses and thermal refurbishment of old build-ings can greatly reduce the heat demand of buildings. By 2050, in accordance with the European energy strategy, the use of fossil fuels in heating systems should be completely substituted by renewable energy. Heat pumps can be widely used, especially in single- and two-family houses. They can also be used in apartment buildings, but there they have to replace the widespread district and local heating supply. In particular, if local heating based on biomass is displaced, there will be additional demand for electricity for heat pumps and direct electrical heating.

The low specific heating energy demand in low-energy houses and refurbished old buildings allows the use of heat pumps with low rated outputs. Direct electric heating systems are also economical to operate in low-energy buildings, even if electricity prices continue to rise in the future.

Electric storage heaters with increased storage capacities and low thermal transmission losses can play a significant role in the expected predominantly renewable energy supply. They can store the surplus energy from fluctuating gen-eration as heat and thus contribute to stabilization of the electrical energy system. This can reduce the need for electrical storage capacities in the electrical energy system.

Storage heaters should therefore no longer be designed as night storage units for a one-day heat storage capacity, as has been the case in the past. Longer peri-ods of lull or surplus of renewable energy require storage capacities of several days. The storage capacity used so far in the non-insulated building stock would already meet these requirements due to better insulation standards of buildings and thus of lower heating energy demand in the future.

References

[Baehr 1989]   Baehr, H.D.: Thermodynamik, 7th edn. Springer (1989)
[Bauphysik 2007] Pech, A., Pöhn, C., Bednar, T., Streicher, W.: Bauphysik. Springer (2007)
[BMWiE 2014]   Sanierungsbedarf im Gebäudebestand, Bundesministerium für Wirtschaft und Energie (2014)
[Bonin 2016]   Bonin, J.: Handbuch Wärmepumpen: Planung und Projektierung. Beuth (2016)
[Cerbe 2005]   Cerbe, G., Wilhelms, G.: Technische Thermodynamik. Carl Hanser (2005)
[dena 2016]   dena-Gebäudereport, Statistiken und Analysen zur Energieeffizienz im Gebäudebestand. Deutsche Energie-Agentur (2016)
[DIN EN 12831]   Heizungsanlagen in Gebäuden – Verfahren zur Berechnung der Norm-Heizlast. Beuth
[EU 2010/31]   Directive 2010/31/EU of the European Parliament and the Council on 19 May 2010 on the energy performance of buildings
[IWU 1996]   Jahresdauerlinien für Niedrigenergiesiedlungen. Institut Wohnen und Umwelt (1996)

[IWU 2010]   Datenbasis Gebäudebestand. Datenerhebung zur energetischen Qualität
und zu den Modernisierungstrends im deutschen Wohngebäudebereich. Institut Wohnen und
Umwelt und Bremer Energie Institut (2010)

[NEEAP-AT 2014]   Erster nationaler Energieeffizienzaktionsplan der Republik Österreich 2014
gemäß Energieeffizienzrichtlinie 2012/27/EU, Anhang B Gebäuderenovierungsstrategie
Österreich

[TUW Ecofys 2015] Bewertung des Potenzials für den Einsatz der hocheffizienten KWK und
effizienter Fernwärme- und Fernkälteversorgung. TU Wien und Ecofys, beauftragt durch:
BMWFW (2015)

[TUW 2018]   Wärmezukunft   2050.   Erfordernisse   und   Konsequenzen   der
Dekarbonisierung von Raumwärme und Warmwasserbereitstellung in Österreich. Studie im
Auftrag von Erneuerbare Energie Österreich (2018)

Efficiency of Mobility

9

9.1 Characteristics of Road Traffic

The transport sector in Germany had a share of 29.5% of the total final energy demand in 2015 [BMWi 2017]. Rail transport has a share of only 2.5%, air transport 13% and inland waterways transport less than 1%. At 84%, road traffic is predominantly responsible for the energy demand of the transport sector. The energy demand of the internal combustion engines or flight turbines have a share of 98% and the electric propulsion represents only 2%. Renewable fuels account for 4% of the propulsion energy. This means, that the share of fossil fuels in the transport sector is 94%.

In the area of road transport, the greatest potential for efficiency lies in the switch to zero emission electromobility with purely electric vehicles or to fuel cell vehicles with electrolysis hydrogen.

Figure 9.1 shows the shares in the energy demand of the individual vehicle types [StatBA 2017]. Passenger car traffic has the largest share with an average mileage per vehicle of 14,000 km/a, followed by heavy goods traffic with a mileage of 62,400 km/a and light commercial vehicles with 23,200 km/a.

The average energy demand of vehicles with gasoline or diesel engines will be given in kWh/100 km in the following, to simplify a comparison in the event of the future switch to electric drives. Figure 9.2 shows the average energy demand of the vehicle types with internal combustion engine. By converting to an electric drive with an accumulator and taking into account a charging efficiency of 80% and a grid efficiency of 95%, 75% of the energy generated by the renewable sources can be used for the electric drive.

Compared to the internal combustion engine, the electric drive only needs about 20% as drive energy, with otherwise identical vehicle characteristics. The 'tank-to-wheel' or 'accumulator-to-wheel' efficiency considered here is to be expanded for the electricity demand, including the charging efficiency of 80%

© Springer Fachmedien Wiesbaden GmbH, part of Springer Nature 2022
G. Brauner, *System Efficiency by Renewable Electricity*,
https://doi.org/10.1007/978-3-658-35138-0_9

Fig. 9.1 Share in final
energy demand by vehicle
type in the road traffic
[StatBA 2017]

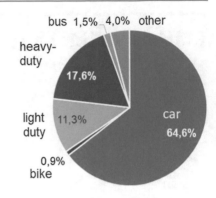

Fig. 9.2 Specific energy
demand of vehicle classes
with internal combustion
engine (blue) [StatBA
2017] and comparison with
efficiency improvement
potentials for electric drives
(red)

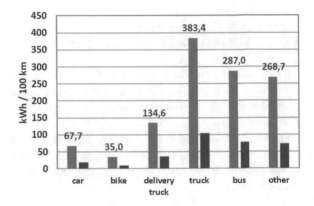

and the network efficiency for the transmission of energy from the regenerative sources. This means that here 28% is required as electrical energy from regenerative sources compared to the fuel requirement from mineral oils. This is shown in Fig. 9.2.

If a similar approach were taken when considering the combustion engine, the energy demand would have to be related to the energy content of the crude oil before it is converted in refineries. Overall, the conversion to electric drives makes it possible to reduce the energy demand in road transport to about one-third of the demand in 2015, which means that the economic and environmentally possible renewable potential is proportionately sufficient for this sector in the efficiency improvement.

This analysis is based on the assumption that the annual mileage of individual vehicle types will remain unchanged. By switching to intermodal concepts with digitization, further increases in efficiency are possible.

Fig. 9.3 Scheme of a pure electric vehicle (BEV/AEV)

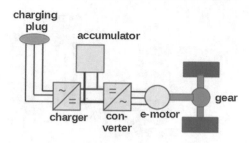

Table 9.1 Charging capacities

Charging current	100 V	230 V	3 × 230 V
10 A	1.0 kW	2.3 kW	6.9 kW
16 A	1.6 kW	3.7 kW	11 kW
32 A	3.2 kW	7.4 kW	22 kW
64 A	6.4 kW	14.7 kW	44 kW

9.2 Types of Electric Vehicles

The first generation of vehicles with electric drives had an electric motor that drives one axle via a differential gear. Figure 9.3 shows the block diagram of such a vehicle. This type is called battery electric vehicle (BEV) or better accumulator electric vehicle (AEV).

The charging electronics can be designed for single-phase or three-phase connection. Table 9.1 shows the possible charging powers.

According to studies in the model region for electromobility in Vorarlberg in the research project VLOTTE [VLOTTE 2010], the mean driving distance per day is 32 km. For these ranges, a daily energy demand of about 4 kWh arises and charging via a normal socket with up to 3.7 kW is therefore sufficient. Higher charging powers are required for long-distance traffic. The trend is towards charging powers of 44 kW, 55 kW and up to 150 kW. This would make the charging electronics in the vehicle very complex and voluminous. Therefore, for high charging powers, the charging electronics are part of the stationary charging devices, and the accumulator in the vehicle is charged directly with direct current.

With the introduction of electromobility, there is still a shortage of charging points and therefore vehicles with a range extenders or plug-in hybrid electric vehicles (PHEV) also used. Vehicles with a range extender or with serial hybrid drive have the schematic concept shown in Fig. 9.4. The drive is electric. A generator for recharging the battery can be switched on via a small internal combustion engine. This provides greater range reliability.

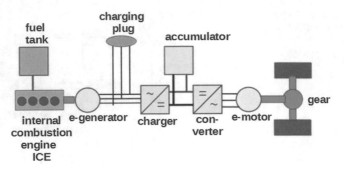

Fig. 9.4 Schematic of a plug-in hybrid electric vehicle (PHEV)

9.3   Energy Demand of Vehicles

The drive power and the energy demand per driven distance are determined in the vehicle by the driving resistance. It is made up of the following components:

Non-reversible components: In this case, the drive energy used is lost.

- **Air resistance** F_L, it is particularly effective at high speeds.
- **Rolling resistance** F_R, it acts at high vehicle weight and bad roads.

Reversible components: The drive energy can be partially recovered or used, for example, by regenerative braking or rolling it out.

- **Gradient resistance** F_S, the absorbed potential energy of the vehicle during uphill travel can be used again to charge the accumulator in the electric vehicle by regenerative breaking during downhill travel.
- **Acceleration resistance** F_B, the absorbed kinetic energy of the vehicle for acceleration can be partially recovered when the vehicle is decelerated by regenerative braking.

Air resistance FL
The air resistance is a drag force that acts on the vehicle and must be overcome by the propulsion power of the engine.

$$F_L = \frac{1}{2}c_w \cdot A \cdot \rho \cdot v^2 \tag{9.1}$$

c_W coefficient of flow resistance
A projected frontal area of the vehicle
ρ air density 1.2 kg/m^3
v velocity in m/s

Streamlined vehicle bodies and small projected frontal areas of the vehicle, as well as spoiler edges at the rear of the vehicle, promote small air resistance. Typical characteristic values for the vehicle classes are shown in Table 9.2. In heavy-duty traffic, the vehicle length and the type of superstructure also determine the aerodynamic drag [FAT 281]. By s spoiler on the tractor or on the trailer and by improving the aerodynamics on the mirrors and at the rear of the trailer, the air resistance can be reduced by up to 12% [FAT 237].

Rolling resistance F_R
Rolling resistance is determined by the condition of the road and the properties of the tires. Smooth road surfaces and tires with low rolling resistance result in low drag forces.

$$F_R = c_R \cdot m_{FZ} \cdot g \tag{9.2}$$

C_R coefficient of rolling resistance
M_{FZ} vehicle weight (empty weight and payload)
g gravitational acceleration 9.81 m/s^2

The rolling resistance coefficient is related to the entire vehicle with four tires and not to the individual wheel. The data from tire manufacturers apply to the individual tire when rolling in the test stand on a smooth iron drum [FAT 304]. The rolling resistance in Eq. (9.2) applies to all tires of a vehicle on the road [FAT 255] and is shown in Table 9.3.

Worn tires have lower rolling resistance but longer braking distances. Also, a design of tires for better grip on wet roads leads to higher rolling resistance. A compromise solution is therefore necessary between driving efficiency and driving safety.

Gradient resistance F_S
The gradient resistance describes the resistance force when the vehicle moves on an inclined plane (Fig. 9.5). This can be described either by the gradient angle α or by the slope gradient p of the road in percent.

Table 9.2 Air resistance parameters of vehicles

	c_W	A	F_L at 100 km/h
		m^2	N
motorcycle (bike)	0.60	0.80	220
small car	0.32	1.80	270
middle class	0.33	2.40	370
off-road vehicle	0.35	2.70	440
delivery trucks	0.42	3.5	570
bus	0.5–0.6	8.7	1700
truck	0.5–0.7	7.0	1950
truck tractor	0.7–1.0	7.3	2900

Table 9.3 Rolling resistance coefficient of road vehicles

Rolling resistance of tires on	Rolling resistance coefficient
Asphalt	0.008–0.015
Cobblestones	0.015–0.030
Earthy road	0.030–0.050
Sand road	0.200–0.400

$$F_S = m_{FZ} \cdot g \cdot \sin\alpha = m_{FZ} \cdot g \cdot \frac{p}{\sqrt{1+p^2}} \qquad (9.3)$$

m_{FZ} vehicle weight (empty weight and payload)
g gravitational acceleration 9.81 m/s²
α pitch angle
p slope in per unit, e.g., 5%: $p=0.05$

The vehicle absorbs potential energy corresponding to the difference in altitude h traveled $E_p = m\,g\,h$. When traveling downhill, this energy, reduced by the rolling resistance loss, can be used to charge the accumulator in the case of electric vehicles. In the case of vehicles with internal combustion engines, it acts partly as a moving force and is partly converted into heat loss via engine braking or special retarders in the case of buses or trucks.

Acceleration resistance F_B.
The acceleration resistance represents a force effect by the mass inertia of the vehicle. The absorbed kinetic energy $E_K = mv^2/2$ can also be used reversibly.

$$F_S = m_{FZ} \cdot b \qquad (9.4)$$

m_{FZ} vehicle weight (empty weight and payload)
b acceleration of the vehicle in m/s²

In electric vehicles, the kinetic energy can be partially recovered by regenerative braking, in vehicles with internal combustion engines by rolling it out.

Fig. 9.5 Incline resistance

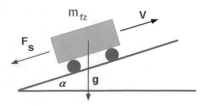

Required drive power of vehicles
Sufficient engine power is necessary to overcome the driving resistance. The performance is determined by the product of the sum of force components of driving resistances and the speed.

$$P_{motor} = [F_L + F_R + F_S + F_B] \cdot \frac{v}{\eta_{motor} \cdot \eta_{gear}} \tag{9.5}$$

$$P_{motor} = \left\{ \frac{1}{2}c_w \cdot \rho \cdot A \cdot v^3 + m_{FZ} \cdot v \cdot [c_R \cdot g + g \cdot sin\alpha + b] \right\} \frac{1}{\eta_m \cdot \eta_g} \tag{9.6}$$

On a level road surface, the gradient forces do not apply. At a constant speed also acceleration forces do not exist. The limiting speed is then essentially determined by the air resistance and the rolling resistance.

The product of the efficiency of the engine and the transmission is in the range of 0.15 to 0.35 for the internal combustion drive. In the following, an average value of 0.25 is used. In case of the electric drive, the overall efficiency of the drive is in the range from 0.7 to 0.8. This means that the electric drive requires only around 30% of the drive energy of an internal combustion drive, with otherwise identical characteristics of the driving resistances. Eq. (9.6) can be used to derive how efficient road transport becomes possible:

- avoid high speeds
- accelerate slowly
- make vehicles streamlined with a small frontal area
- use smooth road surface and tires with low rolling resistance (check air pressure)
- let it roll out with foresight instead of braking
- use recovery the potential energy in the e-mobile through regenerative braking
- recover the acceleration energy through regenerative braking
- use a car with sufficiently large electric motor (used as generator during braking) and converter to be able to fully use the braking energy

9.4 Driving resistance example

What engine power is required for an electric vehicle and a vehicle with a combustion engine to reach a limit speed of 150 km/h?
   Characteristic values of the vehicles:

- $c_w = 0.33$; $c_R = 0.013$; $A = 2.4$ m²; $m_{FZ} = 1400$ kg, no gradient and acceleration.

$$P_{motor} = \left\{ \frac{1}{2}c_w \cdot \rho \cdot A \cdot v^3 + m_{FZ} \cdot v \cdot c_R \cdot g \right\} \frac{1}{\eta_m \cdot \eta_g}$$

Motor power without considering the efficiency of the drives:

$$P_{mot,without} = 41.8 \text{ kW}.$$

Efficiency: Combustion drive $\eta_{motor} \cdot \eta_{gear} = 0.35$, electric drive $\eta_{motor} \cdot \eta_{gear} = 0.75$
primary engine power of the internal combustion engine: $P_{mot,ICM} = 119$ kW
(The primary power is related to the energy of the fuel, the necessary secondary shaft power of the combustion engine is the same as for the electric motor.)
Motor power of the electric drive: $P_{mo, elect} = 56$ kW
Vehicles have greater engine power in order to have sufficient acceleration power for safe overtaking at high speeds. In electric vehicles, high engine power is also beneficial for regenerative braking, since the engine then acts as a generator and the rated output determines the possible braking deceleration with recuperation.

9.5   Energy Demand and Range of Electric Vehicles

The limited ranges and the long charging times compared to the refueling times for internal combustion drives are the main obstacles to the rapid introduction of electric mobility in long-distance transport. The high acceleration, the ability to recuperate the braking energy, the high efficiency of the drive and the possibility to use electricity directly from renewable sources without conversion losses make electromobility appear attractive. The emission-free drive with its high efficiency in stop-and-go traffic during traffic congestions make it appear as a suitable means of transport, especially in urban and suburban traffic. In long-distance traffic, the short range and the need for fast charging still pose partial challenges for the successful and competitive use of electromobility here as well.

The four electric vehicles shown in Table 9.4 are examined with regard to their energy demand and their range.

A rolling resistance of 0.013 for normal tires on asphalt is assumed for all vehicles. The drive efficiency for all electric vehicles is 0.8. Figure 9.6 shows the required electrical drive power as a function of the driving speed according to Eq. 9.6.

Table 9.4 Characteristics values of electric vehicles in 2018

Manufacturer	Type	C_W	Front area A	Weight m_{FZ}	Accumulator capacity
			m^2	kg	kWh
BMW	I3	0.29	2.38	1460	35.0
VW	E-Golf	0.27	2.10	1585	35.8
Renault	ZOE	0.29	2.55	1550	41.0
Smart	fortwo-e	0.37	1.95	1085	17.6

Fig. 9.6 Required drive power of electric vehicles

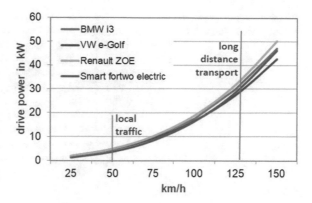

Fig. 9.7 Energy demand as a function of the driving speed

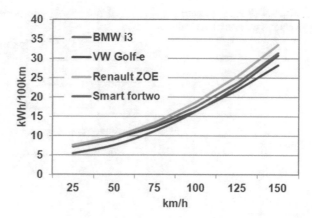

The simulation applies to a level roadway without acceleration. In local traffic, the required drive power is 5 kW; in long-distance transport, 35 kW are required due to the air resistance at 130 km/h.

Figure 9.7 shows the energy demand per 100 km of travel. It increases from around 10 kWh/100 km in local traffic to 30 kWh/100 km in long-distance traffic at higher speed.

The range of the vehicle can be determined from the energy demand together with the capacity of the accumulator. At low speeds in city traffic, the range is high. In long-distance traffic, the range decreases towards 100 km (Fig. 9.8).

From the point of view of the use of the vehicle, the small ranges would be sufficient in city traffic and large values would be desirable in long-distance traffic. The figure shows, that the electric vehicles available even at the beginning of electromobility, have for urban and rural mobility fully sufficient functionalities. In long-distance traffic, larger accumulator capacities and the spread of fast charging devices will also make electromobility attractive in the future.

Fig. 9.8 Range of electric
vehicles according to Table
9.4

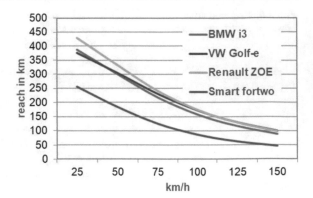

Fig. 9.9 Range of electric
vehicles with auxiliary units

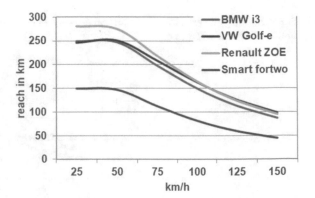

Up to now, the ranges have been determined without taking the auxiliary units into account. In particular, heating and cooling as well as lighting reduce the range. In the electric vehicle, high efficiency is required here. Heating and cooling can be designed efficiently using heat pumps. Furthermore, a significant reduction in the heating energy requirement from the vehicle electrical accumulator is possible with heat accumulators that have been precharged from the grid. The use of LEDs also enables the lighting to be designed efficiently. Figure 9.9 shows the range of electric vehicles if a constant power demand of one kilowatt is assumed for the auxiliary units.

The picture shows that at low driving speeds, the longer driving times mean that there are no advantages in terms of range. At very low driving speeds or at a standstill, the range even decreases due to the auxiliary units being used.

The simulation calculations show that electric mobility requires different vehicle designs for local and long-distance traffic. Small and light vehicles with a short range are sufficient for local traffic. Traditional usage habits, however, assume a universally usable vehicle that is equally suitable for local and long-distance

Table 9.5 Characteristic values of vehicles with internal combustion engine and electric motors

Vehicle type	C_W	Front area A	Weight m_{FZ}	Engine power
		m^2	kg	kW
small truck	0.57	4.29	2880	221
off-road vehicle	0.36–0.37	2.7–3.0	2100–2300	80–280
middle class	0.27	2.19	1500	135
electric vehicle	0.29	2.38	1460	125

Fig. 9.10 Energy demand of combustion drives and electric drives

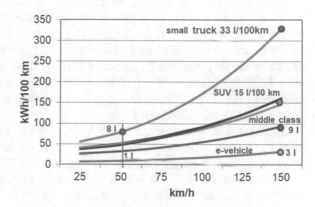

transport. The so-called "range anxiety" is an emotional obstacle to the wide-spread introduction of electric mobility.

9.6   Comparison of Electric and Combustion Drives

The specific energy demand of vehicles with internal combustion engines compared to electric vehicles will be investigated for the various vehicle classes. The vehicles range from small trucks (Hummer H1) to off-road vehicles (SUV) from various manufacturers to mid-class vehicles (e.g., VW Golf), all with combustion engines.

Table 9.5 shows the characteristic values of the vehicles.

Figure 9.10 shows the specific energy demand in kWh/100 km. For comparison, the equivalent fuel requirement in liters per 100 km is shown as the mean value of diesel and gasoline. Diesel has an energy content of 9.7 kWh/l, gasoline 8.6 kWh/l. The electric vehicle is very efficient with an equivalent energy requirement of 1 l per 100 km in city driving and 3 l in long-distance driving at 150 km/h. Off-road vehicles and small truck have very high consumption values, especially at high speed in long distance-traffic. This shows that efficient long-distance transport without emissions is possible through electromobility.

9.7   New European Driving Cycle and WLTP Cycle

The New European Drive Cycle (NEDC) was used to assess the energy demand of vehicles in accordance with Directive 91/44 EEC [NEDC 1991].

The driving cycle consists of four urban and one extra-urban cycle (Fig. 9.11). The urban cycle has a duration of 195 s each and the extra-urban cycle of 400 s. In total the travel time is 1180 s, or just under 20 min. For vehicles with an internal combustion engine, the energy demand is calculated from the exhaust gas values. In electric vehicles, the energy taken from the accumulator is used.

According to the NEDC, this results in low consumption values of 12 kWh/100 km and high ranges of 350 to 400 km for electric vehicles in the Golf-e or BMW i3 class. This does not take into account the energy demand of the auxiliary units. The NEDC strongly emphasizes urban traffic at low speeds and is therefore not very suitable for assessing the specific demand and range of vehicles that are mainly used in long-distance traffic at higher speeds. Here, the specific energy demand can be three times higher and the range can therefore be a third of the values determined with NEDC.

Different driving cycles are in use internationally. In the USA, several driving cycles are used, which are carried out at higher speeds, sometimes on highways with the air conditioning switched on and also at low temperatures.

Instead of the NEDC, the WLTC test cycle is now used internationally. Several cycles are planned, which apply to inner-city traffic, traffic in rural areas for rural inner-city and rural traffic and on highways or expressways. In 2013 the "Worldwide harmonized light duty driving test cycle" [WLTC 2013] was defined, to replace the NEDC starting in 2017. It is defined for three different classes of light vehicles. Class 1 applies to very small vehicles used primarily in developing and emerging countries. The maximum speed of this tests is 64.4 km/h, the test distance is 11.428 km and the test time 1611 s. Class 2 also applies to small vehicles also used in Europe and has a top speed of 123.1 km, a distance of 22.649 km and a duration of 1800 s. Test 3 applies to fleets of passenger cars that are predominantly used in Europe. It is divided into class 3a for vehicles with a maximum

Fig. 9.11 New European
Driving Cycle (NEDC)

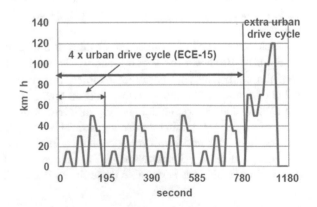

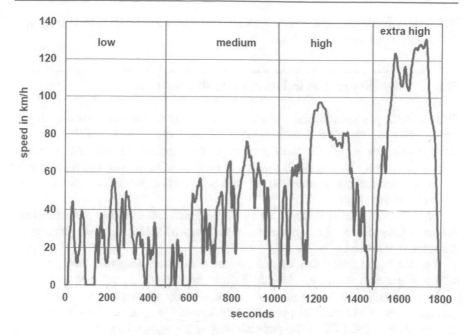

Fig. 9.12 WLTC test cycle for class 3b [WLTC 2013]

Table 9.6 Test cycles in comparison [VDA 2018, WLTC 2013]		NEDC	WLTP 3b
	Start condition	cold start	cold start
	Test temperature	20–30 °C	14–23 °C
	Cycle length	11,000 m	23,250 m
	Cycle time	1180 s/20 min	1800s/30 min
	Speed medium	34 km/h	74 km/h
	Speed maximum	120 km/h	131 km/h
	Acceleration medium	0.39 m/s^2	0.50 m/s^2
	Acceleration maximum	1.04 m/s^2	1.58 m/s^2

speed of less than 120 km/h and class 3b for maximum speeds of 120 km/h or above. Class 3b is the most commonly used test in Europe. The maximum speed is 131.3 km/h, the distance 23.266 km and the duration 1800 s.

The "WLTP Class 3b cycle" with slightly rounded values for Europe is shown in Fig. 9.12 and Table 9.6.

The WLTC test cycle [WLTC 2013] has a longer cycle time, higher acceleration and top speed and requires higher engine power [VDA 2018]. This results in demand values that are up to 25% higher than with the NEDC (Table 9.6). The

WLTP test delivers more realistic demand values than the NEDC and is supported by the EU, USA, Japan and China.

9.8   Mobility Needs in Urban and Rural Areas

Private traffic accounts for the majority of the energy demand in the transport sector. Mobility needs are different in metropolitan areas with very well-developed local public transport (LPT) than in rural or small-town areas with less attractive public transport (PT). The small towns and rural areas have more individual traffic by the car and less public transport by train or bus with lower frequency rates than is possible in the big city.

The different conditions for a big city like Vienna and a rural region with small towns as the province of Lower Austria are in shown Table 9.7 in their characteristic values for mobility.

The high settlements density and the short distances lead to a small number of passenger cars in Vienna. Around 35% of households do not own a car. Due to its large area with low settlement density, Lower Austria has a high passenger car stock of 643 vehicles per 1000 inhabitants. On average, there are 1.5 cars per household, and 87% of the population over the age of 18 years have a driver license.

Under these conditions, local public transport (LPT) with subway, tram and bus as well as pedestrians dominates in metropolitan areas, as shown Fig. 9.13.

In Vienna, 38% of all trips are made in public transport [Wien 2015], in Lower Austria only 13% [Herry 2008], but the car is very important here.

Only short distances are covered by bicycle and on foot in both regions, as shown in Fig. 9.14. In Lower Austria trips on foot have a mean distance of 1.4 km and amount to a maximum of 4 km to 5 km. Trips on bicycle have a mean distance of 2.5 km in Lower Austria and amount to 10 to 20 km. In the metropolitan area, the distances are shorter. Longer distances are covered by car (motorized individual transport MIT) and public transport (PT) by train and bus in lower Austria, which occur most frequently up to 50 km. The average distance per trip is significantly greater in rural regions than in urban areas.

The average travel times are shown in Fig. 9.15. They are on average 30 min per one-way distance. In public transport by bus and train, the average travel times

Table 9.7 Mobility parameters for Vienna and Lower Austria (Statistics Austria)

		Vienna	Lower Austria
Area	km^2	414	19,186
Inhabitants	PE	1,840,000	1,670,000
Settlement density	PE/km^2	4450	87
Cars per 1000 inhabitants (2017)		381	643

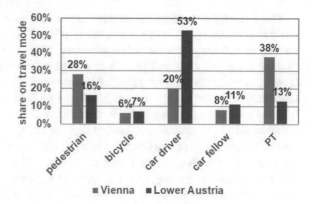

Fig. 9.13 Mobility by mode of transport

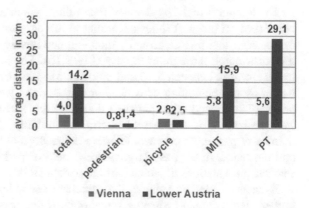

Fig. 9.14 Average distances by mode of transport

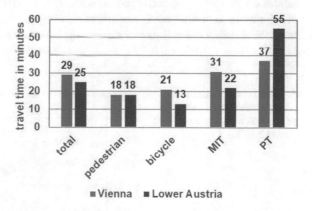

Fig. 9.15 Average travel times by to mode of transport

is up to 55 min. This is due to the intermodality with travel times to the bus stop or train station and waiting times until departure. There are between 2.9 and 3.5 trips per person per day.

These results show that the electric car can be widely used in the future, especially in rural and small-town areas, because it is immediately available and avoids start-up or waiting times for public transport. Here, in particular, it can replace the passenger car with internal combustion engine in a renewable and emission-free manner. It will also compete with public transport for short distances. In long-distance traffic, on the other hand, the electric vehicle as a feeder can enlarge the area from which intermodal mobility is possible. This will give the railways in rural areas incentives to restructure. Station density can then be reduced because of the greater feeder range of the electric vehicles, and the railroads can be developed in the direction of high-speed rail with fewer stops. When electric cars become suitable for automated driving, they can serve as a feeder of all road users—including young and old with no driving license—and revolutionize the suburban mobility through automated intermodal mobility.

The temporal mobility demand determines which mobility concepts are possible. Based on the start times according to the purpose of the trip [Herry 2008], it can be seen that private errands, shopping and leisure time, result in a traffic volume distributed throughout the day (Fig. 9.16). The highest traffic volume occurs in the morning between 5:30 and 8:00 am. Morning traffic is characterized in particular by the beginning of education and work as well as bringing and fetching people. The related travel times or departure times of bus and train determine the synchronization. In the case of intermodal mobility, this means that a very large number of passenger cars would arrive at the stops at the same time and require a parking space. In the past, large parking areas or parking garages were created at selected train stations in which cars are parked all day.

Even as an electric vehicle, the passenger car is less efficient here in terms of higher energy demand when occupied only by one person and because of the inefficient use of parking spaces. The automatically driving minibus can be used more efficiently in suburban areas and, through digitization, can enable more demand-oriented mobility via cell phone requests (see Sect. 9.8.2).

Fig. 9.16 Start times of the routes according to route purpose on working days [Herry 2008]

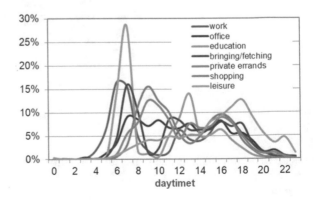

In the metropolitan mobility, the simultaneity of the mobility needs in the "rush hour" leads to an increased demand for transport capacity in public transport. Commuters with cars experience traffic jams in the access roads.

9.9   Efficiency Potential in the Transport Sector

9.9.1   Efficiency Through Technology

The technological efficiency potential of electromobility results from a complete conversion from the internal combustion engine to the electric drive. The first column of Table 9.8 shows the share of vehicle types in the energy demand in the mobility sector. The second column shows the share on the total primary energy demand in 2015. In most industrialized nations, the mobility sector has a share of around 30% of the total primary energy demand. Only percentages are shown in the table, as the results are then more generally valid.

The improvement in energy efficiency in the mobility sector compared to the year 2015 is to be made in relation to the average energy demand of passenger cars in 2015 [STABA carried out in 2017]. A passenger car with an internal combustion engine had an average energy demand of 67.8 kWh/100 km. An electric vehicle would get by with an energy demand of 14 kW/100 km.

There types of electric vehicles are assumed in 2050:

- Accumulator Electric Vehicle (AEV), is also known as Battery Electric Vehicle (BEV). Since the primary energy demand is considered, the efficiency of the battery in the charge–discharge cycle of around 80% and the grid efficiency of 96% for the transmission of renewable energy to the charging point must

Table 9.8 Technological efficiency potential of electromobility

	Energy sector share (%)	Primary energy share (%)	AEV/ BEV- share potential (%)	FCEV share (%)	Primary energy AEV/ BEV (%)	Primary energy FCEV (%)	Primary energy AEV&FCEV (%)
Car	64.4	19.38	90	10	4.70	1.34	6.03
Motorcycle	0.9	0.28	100	0	0.08	0.00	0.08
Light duty	11.3	3.40	60	40	0.55	0.94	1.49
Heavy duty	17.6	5.29	10	90	0.14	3.28	3.43
Bus	1.5	0.44	10	90	0.01	0.27	0.28
Other	4.0	1.21	90	10	0.29	0.08	0.38
Total	100	30			5.77	5.91	11.68

be taken into account. This results in an increased primary electricity requirement of 18.2 kWh/100 km. In general, the ratio of 18.2/67.8 = 0.2693 is used as the efficiency factor for all vehicle types with this type of drive for the ratio of primary electrical energy to the energy of mineral fuels when switching to electromobility.

- Fuel Cell Electric Vehicle (FVEV). Electrical drives with hydrogen as the energy carrier and fuel cell are advantageous in passenger and freight transport because they have short refueling times and do not require rapid charging of accumulators. However, they have a higher primary electricity requirement than the pure electric drives. Hydrogen electrolysis has an efficiency of about 60% and the fuel cell 50%. Converting a passenger car to fuel cell technology, results in a higher efficiency factor because the primary electricity demand is 46.7 kWh/100 km and the pure internal combustion engine drive hat 67.7 kWh/100 km. The efficiency factor of the fuel cell vehicle is therefore 46.7/67.7 = 0.6895.

- Electric drive with trolley wires. The heavy-duty transport sector can obtain the drive energy from two overhead trolley wires on specially prepared routes, similar to what has been in use for a long time with the trolleybus. This eliminates the need for frequent rapid charging such as electric drives with accumulators. Compared to an electric drive with fuel cells, the overall efficiency is also higher, since only the grid losses occur when during energy supply. Such a project of ecologically compatible heavy-duty traffic was investigated for the Alpine transit between Austria and Italy [TUW 2000 and TUW 2003] with trolley wires and vehicles with electronic distance control. In order to avoid reloading the trucks, hybrid drives should be used either with accumulators or with fuel cells, which allow continued driving when leaving the electrified routes. In the study of the efficiency potentials, O-trucks can be treated as accumulator electric drives.

In the Table 9.8 is assumed that in 2050, 90% of heavy-duty vehicles and buses will be converted to fuel cell propulsion, 40% of light -duty vehicles, and only 10% of passenger cars and other vehicles.

As shown in the last column of Table 9.8, the total primary electricity demand has a value of 11.68% compared to 30% for internal combustion propulsion, based on the primary energy demand of 2015. This reduces the primary energy demand to 39% of its value of 2015, and the transport sector achieves the saving targets set by the available renewable potential. Due to of the lower efficiency in hydrogen production and electricity conversion in the fuel cell, the primary electricity demand for the FCEV stock is 5.91% higher than for AEV/BEV stock at 5.77%.

Further technological efficiency potentials are possible through small and lightweight electric vehicles with low c_w values, small frontal areas, low rolling resistance and efficient auxiliary units as well as efficient lighting.

In the previous considerations, the same traffic volume was assumed in the future. In the past decade, the road traffic has increased by 10% and rail freight by

5% [BMVIT 2012]. Passenger freight transport as motorized individual transport and as public transport has also experienced an overall increase of 10%. Therefore, the technological changeover from the internal combustion engine to the electric drive presented here alone will not achieve the efficiency targets. Additional measures such as freight logistics, intermodal transport systems, automated mobility, car-sharing and changed user behavior are necessary.

9.9.2   Efficiency Through Organization and Digitization

In rural areas, there is great potential for increasing the efficiency of individual transport through organization and digitization. Digitization is also a technology that opens up many new uses.

Car-sharing represents the first form of digitization. The cell phone is used for communication of the user, he can see the location of the vehicle, find out fuel or charging status, and open and use the vehicle by means of an RFID chip or via the cell phone. The distance traveled can be recorded using the GPS system in the vehicle and the user's account can be debited automatically. Such car-sharing-systems as *car-2-go* or *drive now* are operating successfully.

Car-sharing is currently only possible within large cities, where there is already a well-developed public transport system. In rural areas, car-sharing is not economically feasible with today's technology, as it requires a high population density. If, for example, a vehicle is used in rural areas in a sparsely populated area and is parked somewhere, it can take a long time to find a new user. The vehicles therefore have a low usage rate and long idle times, which is why car-sharing is hardly widespread here.

Automated driving can revolutionize car-sharing in the future in such a way that it can also be used in rural areas and no driver's license is required for the person using it. The vehicles would then be able to drive independently to the place of order upon request via a cell phone, pick up the person and drive automatically to the destination. After use, it would automatically return to the next customer or to stand locations, where it can be charged and from which there is a short travel time to a likely mobility need.

Public transport can also be reorganized through automated driving. Figure 9.17 shows a concept, which uses automated minibuses.

The minibuses drive in the peripheral zones on demand-oriented flexible routes and pick-up passengers. In more densely populated areas, the vehicles can be electronically coupled as a minibus train and thus increase the transport capacity and at the same time use the traffic area more effectively due to the short distance between them. Individual minibuses can also separate themselves from the train in order to drive to specific destinations. This will be communicated to the passengers and a transfer to other minibuses is made possible. Individual minibuses in trains can also serve as a feeder to the train station, educational centers, or workplaces. This enables demand-oriented and automated passenger transport, with

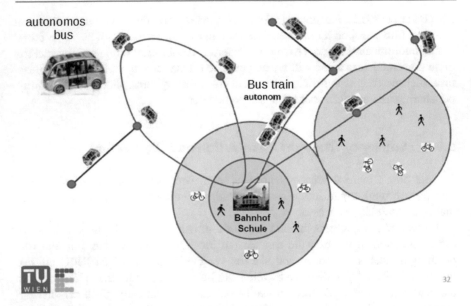

Fig. 9.17   Automated local public passenger transport

variable transport capacity and acceptable cycle times even in the sparsely populated peripheral zones. This is just one example of the many possibilities for digitization and automation of public transport.

Possibilities and limitations of automated driving
Automated driving requires a high level of traffic safety. Preliminary stages for this are driver assistance systems that are already available:

- ABS anti-lock braking system (brake assistant)
- TPM Tire Pressure Monitoring (tire pressure monitoring)
- ESC Electronic Stability Control (driving stabilization)
- AEB Automated Emergency Breaking (emergency braking system)
- ESS Emergency Stop System (emergency braking support)
- IPAS Intelligent Parking Assist System (parking aid)
- LDP Lane Departure Prevention (Lane Keeping Assist)

With the driver assistance systems, the driver participates in the traffic situation (in the loop) and can intervene at any time. He is only supported by the assistance systems.

In fully automated driving, computers in the vehicle take full control. They use sensors with ultrasound for distance, radar, high-resolution cameras and laser scanners for three-dimensional location of the course of the road and the surrounding traffic. The driver does not take an active part in traffic—he is "out of the loop". If the automation system fails, the person must intervene briefly to corrective the situation. However, this is hardly possible at high speeds because of the short reaction times required.

It therefore appears realistic to initially limit automated driving to the range of low speeds, for example to allow a speed of 50 km/h. At higher speeds, for example up to 120 km/h, the person has control of the vehicle and is only supported by assistance systems. At high speeds, the human takes full control. This seems appropriate at the current state of development of automated driving. The limit of automated driving may shift to higher speeds in the future due to new developments.

9.9.3 Efficiency Through User Behavior

Increasing efficiency through user behavior will be of increasing importance in the future. Small distances of up to about 2 km can be covered by walking. Longer distances of up to 5 km can be covered by bicycle, e-bike or cargo bike.

If SUVs, which currently account for about 20% of the vehicle population, are replaced by small electric vehicles, the energy demand in this segment can be roughly halved.

By forming car pools, considerable efficiency potentials can also be activated in commuter and educational traffic.

Using public transport or local public transport instead of your own vehicle results in significant savings. Intermodal use of transport can also improve efficiency and reduce emissions. Existing transport infrastructures with higher numbers of passengers can then be used more efficiently.

Finally, urban planning with the expansion of residential buildings in cities can reduce the volume of traffic.

9.10 Summary

In the transport sector, the switch to electromobility is a necessity to enable an efficient renewable energy supply that can manage with the limited regenerative generation potential. The small, efficient electric vehicle is already available today. The necessary charging infrastructures are being set up in many countries.

The electric vehicle with fuel cell and hydrogen with electrolysis by means of renewable energy will be important in the future in passenger and freight transport, because accumulator charging leads to long waiting times, which are avoided when refueling with hydrogen.

Overall, the primary energy demand in the transport sector can be reduced to 40% of today's demand by switching to electromobility. This would enable the transport sector to achieve the efficiency targets specified on the basis on the potentials. Due to the probable increases in traffic volumes, this target will be missed and additional measures are required. These are the digitization and automation of individual transport and changes in user behavior.

Digitalization and automated driving enable new local transport concepts with high acceptance in small towns and suburban areas. This can further increase efficiency.

By changing user behavior with the expansion of footpaths, use of bicycles, a lower volume of traffic in public transport is possible. By avoiding off-road vehicles and using small and efficient electric vehicles, higher energy efficiency and reduced need for parking spaces are possible.

References

[BMVIT 2012]   Gesamtverkehrsplan für Österreich. Bundesministerium für Verkehr, Innovation und Technologie (2012)
[BMWi 2017]   Energieeffizienz in Zahlen. Bundesministerium für Wirtschaft und Energie (BMWi), Berlin (2017)
[StatBA 2017]   Transportleistungen und Energieverbrauch im Straßenverkehr 2006–2015. Statistisches Bundesamt Deutschland (2017)
[FAT 237]Verbrauchsreduktion an Nutzfahrzeugkombinationen durch aerodynamische Maßnahmen. Forschungsvereinigung Automobiltechnik, Band FAT 237 (2011)
[FAT 255]Untersuchung des Rollwiderstandes von Nutzfahrzeugen auf echten Straßen. Forschungsvereinigung Automobiltechnik, Band FAT 255 (2013).
[FAT 281]Schwere Nutzfahrzeugkonfigurationen unter Einfluss realitätsnaher Anströmungen. Forschungsvereinigung Automobiltechnik, Band FAT 281 (2015)
[FAT 304]Der Rollwiderstand von Nutzfahrzeugen unter realen Umgebungsbedingungen. Forschungsvereinigung Automobiltechnik, Band FAT 304 (2018)
[Herry 2008]   Mobilität in NÖ—Ergebnisse der landesweiten Mobilitätsbefragung 2008. Amt der NÖ Landesregierung, Abteilung Gesamtverkehrsangelegenheiten.
[NEDC 1991]   Directive 91/441/EEC, Council Directive of 26 June 1991 amending Directive 70/220/EEC … relating to measures to be taken against pollution by emissions from motor vehicles.
[TUW 2000]   Lenz, P., Brauner, G., Litzka, J., Pucher, E.: LKW-Alpentransit elektrisch? VDI-Kongress: „Innovative Fahrzeugantriebe". 26./27. Oktober 2000, Dresden. VDI Berichte Nr. 1565:605–623 (2000).
[TUW 2003]   Brauner, G., Lenz, H.-P., Litzka, J., Pucher, E.: „Ökologisch verträglicher Schwerlasttransit in Österreich". Bundesministerium für Verkehr, Innovation und Technologie BMVIT, Schriftenreihe: Forschungsarbeiten aus dem Verkehrswesen, Band 131, Wien (2003)
[WLTC 2013]   Development of a Worldwide harmonized Light duty driving Test Cycle (WLTC). Technical Report UN/ECE/WP.29/GRPE/WLTP-IG (2013).
[VLOTTE 2010]   VLOTTE—Elektrisch Mobil. Vorarlberg Modellregion der Elektromobilität 2010.Gefördert vom Klima- und Energiefonds Österreich. Begleitforschung der TU Wien. Endbericht (2010)
[VDA 2018]   WLTP—neues Testverfahren weltweit am Start. Fragen und Antworten zur Umstellung von NEFZ auf WLTP. Verband der Automobilindustrie (VDA) (2018)
[Wien 2015]   Zu Fuß gehen in Wien. Vertiefte Auswertung des Mobilitätsverhaltens der Wiener Bevölkerung für das zu Fuß gehen. Omnitrend GmbH (2015)

Efficiency Through Sector Coupling

10

10.1 Sector Coupling

Renewable energy supply is dominated by energy sources whose characteristics consist in the conversion of volatile environmental energy into electricity. The variable supply leads to periods of high overgeneration and periods of prolonged generation shortage. Short-term storage of electricity by pumped storage and stationary or mobile accumulators can be provided only partially in sufficient potential with good profitability.

So far, the methods for the flexible use of renewable surplus energy are insufficiently developed. Due to the insufficient capacities of the transmission and distribution grids and the insufficient flexibility of the loads, the frequency of curtailment of wind energy or the shutdown of PV plants is increasing.

The use of renewable surplus energy through coupling to other sectors of energy application represents a possibility in the future to use this energy directly with high degrees of efficiency or to store it in a converted form of energy in the short and long term.

Sector couplings are possible in many ways and are referred to as **Power-to-X technologies**. They can be subdivided according to both forms: the energy converted or the purposes for which they are used.

Coupling by energy sector (forms of energy)
- **Power-to-Heat (P2H)**
  Here, heat is generated either by direct conversion from electricity via a heating rod or an electrode boiler, or indirectly by heat pumps, which deliver a multiple of the electricity used as heating energy with a coefficient of performance of 3 to 5.
- **Power-to-Gas (P2G)**
  Power-to-gas is used to produce renewable gas (hydrogen) by electrolysis from renewable electricity. Hydrogen can be stored as a gas under high pressure.

Furthermore, it can be fed into existing natural gas networks up to a share of around 10 %. Together with carbon dioxide, synthesis to synthetic methane is possible. This is also known as power-to-syngas.

- **Power-to-Liquid (P2L)**
This includes various processes for the production of liquid hydrocarbons. On the one hand, there is the liquefaction of gases in order to reduce the volume and simplify transport. On the other hand, these are synthesis processes in which long-chain hydrocarbons are obtained from electrolytic hydrogen.

Coupling by purpose
- **Power-to-Power (P2P)**
Here, electrical energy is stored or used to generate an energy carrier and then converted back into electricity via an energy converter. When energy is stored in mechanical pumped storage or electrochemical accumulators, the overall power-to-power efficiency is around 80 %. When converted into hydrogen and converted back into electricity using fuel cells, the overall efficiency is around 30 to 40 %. When converted to synthetic methane and converted back to electricity in a combined cycle power plant, the efficiency is 25 to 30%. When assessing the applicability of the different variants, it is not only the efficiency that is decisive, but other factors are also important. For example, in mobility with hydrogen, the time-saving refueling of electric vehicles with fuel cells is an advantage over the rapid charging of accumulators in electric vehicles.
- **Power-to-Mobility (P2M)**
Electric vehicles with accumulators represent a possibility of flexible use of electricity. They can flexibly adapt the charging behavior to the generation situation in the renewable energy supply. Further possibilities are a coupling to power-to-gas with hydrogen for the drive with fuel cells. Power-to-gas with synthetic methane for use with the internal combustion engine is likely to be of minor importance in the future because of poor efficiency and exhaust gases. This is probably also true for the coupling to power-to-liquid.
- **Power-to-Fuel (P2F)**
Power-to-fuel is a collective term for power-to-gas and power-to-liquid and describes the production of gaseous or liquid fuels such as synthetic natural gas (SNG), methanol, and gasoline from electrolysis hydrogen by synthesis with carbon dioxide.

Power-to-X offers a wide range of options for sustainable energy supply

- With power-to-gas and power-to-liquid, a coupling with the gas sector, the mobility sector and the heating sector is possible. Long-term storage with high energy densities in former oil and gas storage sites without self-discharge as with electrochemical storage is another option.
- Power-to-Heat is a cost-effective and flexible way to use renewable surplus energy in the heating sector of buildings and industry.

10.2   Heating Sector – Power-To-Heat

In the field of low-temperature heat, the heat pump represents a highly efficient coupling of the electrical grid to building heating (see Chap. 8). Due to the large thermal storage time constants of buildings, intermittent operation with longer breaks is also possible, corresponding to the volatile nature of the renewable surplus energy.

In the industrial sector, high-temperature electrical heat, for example in the electric arc furnace or in the conductive sintering processes, has long been a link to the heating sector and replaced fossil fuels. Here, too, flexibility is possible – albeit to a limited extent – according to the supply.

In existing heating networks, electrode boilers or heating rods can supplement the existing heat supply from cogeneration plants. There is also a great potential for cogeneration in industry, especially in the sectors of chemicals and paper.

Electrode boilers for hot water or steam generation
Electrode boilers contain three blade-like electrodes that can be connected directly to the medium-voltage network without a transformer (Fig. 10.1). Because of the low investment costs of 150–350 €/kW, they are widely used in district heating networks and for industrial process heat generation.

The water is pretreated and has a low conductivity of 0.001 to 0.005 S/m. The level under the electrodes can be changed to control the output. This allows a wide range of power control. Via a circulation pump, the heated water enters a heat exchanger for extraction of district or process heat. At water temperatures above the boiling point, nitrogen can be used to generate a pressure that prevents boiling.

Electrodes boilers can also be used in district heating networks as a reserve boiler or as a quick-start boiler for peak loads. Table 10.1 shows the technical characteristics of electrode boilers for hot water or superheated steam generation.

Electric boiler
Electric boilers contain heating rots and are suitable for smaller outputs in the range from 10 kW to 15 MW. Like the electrode boilers, they can be designed to generate hot water or steam. The investment costs range from 75 and 200 €/kW.

Fig. 10.1 Electrode boiler
for hot water production
[Parat principle]

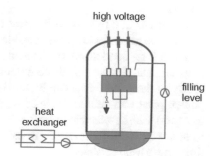

Table 10.1 Characteristic values of electrode boilers

Thermal performance	1–90 MW
Nominal voltage	5–22 kV
Pressure	0.5–55 bar
Temperature	80–240 °C
Efficiency	>99%
Cold start time	<5 min
Load change 0–100%	30 s
Control range	0–100%
Boiler volume	20–40 m^3
Height	5–8 m
Investment costs	125–350 €/kW

Flexibilization of district heating networks with electrode the boilers

District heating networks with cogeneration represent central urban large-scale plants. Combined cycle gas and steam power plants with unit capacities in the range of 50 to 400 MW are still needed, since longer periods without sufficient renewable supply cannot be supplied with storage systems at the high grid loads in industrialized countries. Electrode boilers can be used to feed larger heat outputs in the range of 5 to 90 MW from renewable surplus generation into district heating networks with high efficiency close to 100%.

District heating networks generate their heat demand predominantly from natural gas. Combined cycle power plants can achieve high overall efficiencies of up to about 90% for conversion to electricity and simultaneous extraction of heating energy. The heat is taken from the steam turbine of the combined cycle process. Figure 10.2 shows a steam turbine with two heating condensers as an extraction tap and as a back pressure outlet at the end of the turbine. The extraction tap has the effect that the electrical output power of the generator decreases when steam is extracted [enpros 2014]. With the backpressure extraction at the end as backpressure tapping, the electrical power increases proportionally to the steam extraction and thus to the heating energy (Fig. 10.3). By varying the tapping flows from both taps, the electricity and heat generation can be better adapted to the needs of these sectors.

By means of a bypass with a heat exchanger for connecting an electrode boiler, heat can be generated from renewable surplus electrical power. The heat extraction at the heating condensers of the steam turbine can then be reduced accordingly. If cogeneration plants with these extensions are located near wind farms, the electricity of the wind farms can be used directly and as heat and the combined cycle power plant can reduce its electrical and heat output accordingly. Long-range transport via the electricity grid with possible grid overloads can be so avoided. In this way, fossil fuels can be saved by using renewable energy, that would sometimes have been depleted.

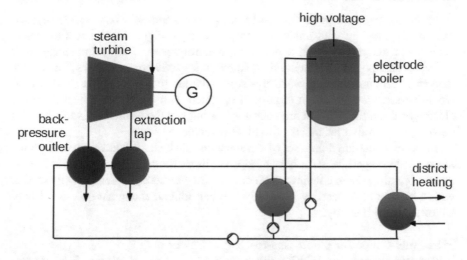

Fig. 10.2   Electrodes boiler in a combined heat and power plant

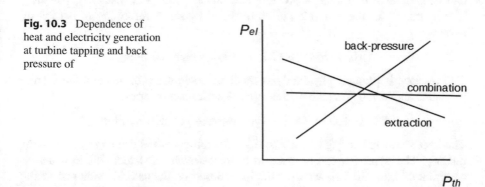

Fig. 10.3   Dependence of heat and electricity generation at turbine tapping and back pressure of

For smaller heating capacities, less expensive electric boilers can be looped directly into the bypass as instantaneous heater without the need for an intermediate heat exchanger. This is a cost-effective solution for smaller local heating networks.

Fuel cell power plant: Power-to-Heat & Power

In the future, decentralized sector coupling of electricity and heat will be possible using fuel cell CHP systems. They can be installed in buildings and dimensioned for electrical power in the range from 1 kW to about 100 kW. In the initial phase of this new technology, natural gas is used as fuel, since the development of small electrolysers for hydrogen production has not yet been completed and the inexpensive renewable surplus energy is only gradually becoming available on the market. Fuel cells powered by natural gas can generate electricity and heat and thus enable buildings to be partially self-supplied.

In the future, sustainable fuel cell mini-power plants will be possible that are exclusively supplied with renewable surplus energy. Figure 10.4 shows the overview block diagram of such a plant. An electrolyser is used to split water into hydrogen and oxygen, which are then stored in separate pressure tanks. Since pure oxygen is produced during electrolysis, an alkaline fuel cell can be used, which would be sensitive to carbon dioxide if operated with air. If waste heat from the electrolyser and the fuel cell are used for heating and the electricity is fed into the power grid, overall efficiencies of to 90% are achieved.

Because of the large number of components, such small fuel cell power plants are not yet competitive with heat pumps or direct heating. Because of their ability to store and generate their own electricity, they represent decentralized systems with a high level of security of supply and can find an economic end use if they are evaluated differently.

Micro fuel cell power plant: Gas-to-Heat & Power
With synthetic methane (SNG), obtained in a power-to-gas process from renewable surplus energy, with biogas or with natural gas, electricity and heat for houses can be generated in a micro-CHP fuel cell power plant [VIT VALOR], [Elcore]. The hydrogen for the fuel cell can be obtained by steam reforming of natural or synthetic gas.

$$CH_4 + H_2O \rightarrow CO + 3H_2 \quad \text{steam reforming}$$

The reaction is promoted by catalysts made of noble metal or nickel at high temperatures above 700 °C and low pressures. A shift reaction arises

$$CO + H_2O \rightarrow CO_2 + H_2 \quad \text{water} - \text{gas shift reaction}$$

The heat generated during the exothermic reforming process can be used for room heating. The hydrogen can be used in a downstream PEM fuel cell to generate electricity. Figure 10.5 shows the principle. Based on the calorific value of methane, the thermal efficiency for heat generation is around 50% and the electrical efficiency for electricity generation is 38%. The mean combined overall efficiency is 86%.

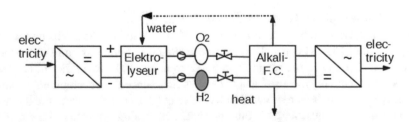

Fig. 10.4 Decentralized electrolysis with gas storage and fuel cell generator

Fig. 10.5 Micro-CHP with methane reformer and PEM fuel cell

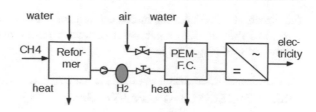

Fig. 10.6 Flexibilization of an industrial process heat network with a heat pump

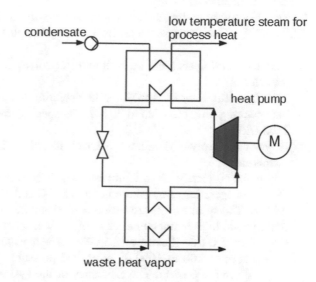

Flexibilization of industrial process heat generation with heat pump

In the industrial sector, process heat with a higher temperature level can be generated by using waste heat from production processes.

Figure 10.6 shows such a process, in which waste heat vapor is condensed in a heat exchanger at the end of the production process [PFI 2013]. The water recovered in this way can be reused for the process. A heat pump is used to convert production condensate into low-temperature steam. By of a downstream compression turbine, superheated steam could also be generated from it. The drive energy for the heat pump is powered by electric motors, which means that the process can be run with renewable electricity.

Control and balancing energy with power-to-heat systems

Primary and secondary control of the electricity transmission grid have time ranges for the provision of control power of minutes for the primary control and about 30 min for the secondary control. Because of the storage capacity in the boilers and pipeline systems of the district heating networks, they can participate in the balancing energy market by changing their electricity supply. Power-to-heat plants with flexibly deployable electrode boilers can offer control energy in the direction of reduced or increased demand, starting from an average reference band, and thus act like a generator or load from the perspective of the secondary controlled grid (Fig. 10.7).

10.3   Gas Sector – Power-To-Gas

Hydrogen electrolysis
The electrolysis to hydrogen can be carried out with the following three processes [ZSW 2017], which represent a reversal of the principle of fuel cells.

- Alkaline electrolysis (AEL) with liquid electrolyte as potash or sodium hydroxide solution.
  The plant efficiency is 51 to79 %, the operating temperature is 60 to 80 °C and the operating pressure is up to 50 bar. The specific investment costs are 1000 to 1200 €/kW.
- Polymer Electrolyte Membrane Electrolysis (PEMEL) with proton conducting membrane.
  The plant efficiency of this process, which is still in development, is 47 to 79%. The operating temperature is 50 to 80 °C and the operating pressure up to 350 bar. The system costs are between 1500 and 2300 €/kW.
- Solid Oxide high temperature Electrolysis with ceramic ion-conducting membrane (Solid Oxide Electrolysis, SOEL). Water vapor is heated to a very high temperature of 700 to 1000 °C at normal pressure. The system costs are still very high at 2500 €/kW. The efficiency of the system is still unknown as this process is only being developed in the laboratory.

Hydrogen electrolysis is suitable for fluctuating partial load operation. The dynamic behavior is determined by the downstream components such as gas separators, liquid circuit and pressure regulator. To compress the hydrogen in the storage tank to 200 to 800 bar, an energy loss of 8 to 13% occurs. In the electrical

Fig. 10.7 Secondary control by absorbing excess power through an electrode boiler

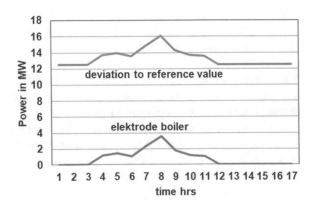

Table 10.2 Characteristic values of electrolysis processes [ZSW 2018]

	AEL	PEMEL	SOEL
System efficiency	51–79%	47–79%	
Operating temperature	60–80 °C	50–80 °C	700–1000 °C
Max. Pressure	<50 bar	<350 bar	1 bar
Partial load capability	20–40%	~10%	
Investment costs	1000–1200 €/kW	1500–2000 €/kW	2500 €/kW

Table 10.3 Characteristic values of synthetic methanation [ZSW 2018]

Temperature range	200–700 °C
Pressure	4–80 bar
Efficiency	70–95%
Specific investments	600–1000 €/kg CH_4

system, consisting of transformer and rectifier, losses of 5% occur. The average overall power-to-gas efficiency of hydrogen electrolysis is around 51 to 79% (Table 10.2).

Hydrogen can be converted back into electricity in fuel cells with efficiencies of 50 to 60%. This enables a coupling to the P2G, P2M and P2P sectors.

Catalytic methanation, CH_4 from CO_2/H_2

For the synthetic production of methane (CH_4), electrolytically obtained hydrogen can be converted into methane with carbon dioxide under high pressures and temperatures (Table 10.3).

Suitable sources with a high CO_2 potential are thermal power plants and the steel and cement industries with CO_2 concentrations of 10–20%, biogas and sewage gas plants and chemical plants with ammonia and ethylene oxide synthesis with concentrations up to 100% [ZSW 2018].

Methane can be stored in a chemically stable manner in exploited natural gas reservoirs. The volumes of these deposits can accommodate equivalent gas energy to a primary energy demand of several years. The synthesis gas can be converted back into electricity in combined cycle power plants with an efficiency of up to 62%. Overall, this results in a power-to-gas-to-power efficiency of around 30%.

10.4 Mobility Sector—Power-To-Mobility

The mobility sector in Germany and Austria accounts for around 30% of the total energy demand. Currently (2018) in this sector 94% from fossil energy is predominately used in form of motor gasoline and diesel fuel and 0.3% from natural gas.

Electricity and renewable fuels only have a share of 5.5%. The share of electricity is currently mainly used for electric railways in long-distance and local traffic.

By converting mobility with a combustion engine to an electric drive, the energy demand can be reduced to at least 30% of the previous demand. This corresponds to a reduction of the energy demand in this sector by 70%, which should be achieved by each sector in accordance with the long-term energy strategy for renewable energy supply in order not to exceed the potential limits.

Three types of sector coupling are possible for the future development of sustainable mobility:

- Coupling to the electricity sector via the electric vehicle with charging strategies to improve the possibilities for using the volatile renewable supply. Another variant is the direct use of electricity in rail vehicles and, in the future, in the transport sector by trucks with trolley line supply.
- Coupling to the sector power-to-gas for vehicles with hydrogen tanks and fuel cells for generating electricity for propulsion. This type of application is particularly important in long-distance transport of cars and trucks because hydrogen refueling stations allow a similar short charging time as with gasoline and diesel fuels. The efficiency of the power-to-tank hydrogen infrastructure is around 60%.
- Coupling to the sector power-to-gas based on synthetic methane. The methane can be used as drive energy in the combustion engine. Since methane synthesis from electrolysis hydrogen has an efficiency of about 35 to 45% and the internal combustion engine about 15 to 20%, the overall efficiency is less than 10%. Furthermore, exhaust gases with nitrogen oxides are produced. Overall, therefore, this type of sector coupling does not appear to be very promising.

Electric vehicle
The main application of electric vehicles is in commuter traffic and suburban private transport. The daily driving distance is around 35 km. This corresponds to a specific energy demand "accumulator-to-wheel" of 14 kWh/100 km and a daily energy demand of about 5 kWh. The losses during charging and discharging are around 20%. A charge for the daily demand is possible within 1.5 h at a normal power socket. Fast charging is necessary for long-distance traffic.

Figure 10.8 shows the charging time for a distance of 100 km as a function of the charging power. For long-distance traffic, the specific energy demand increases to 20 to 100 kWh/100 km due to the higher air resistance or bigger vehicles for transport.

Fuel cell vehicle
Vehicles with electric drive and fuel cell have a high-pressure hydrogen storage that can be recharged in about five minutes. The driving power of trucks is in the range of 400 kW and, according to Table 10.4, the specific energy demand

Fig. 10.8 Fast charging devices

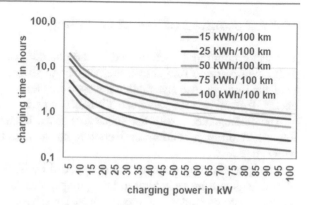

for purely electric drive is 28 kWh/100 km for light goods vehicles and 80 to 75 kWh/100 km for trucks and semitrailer tractors, respectively [DESTATIS 2017]. With a fast-charging device with a power of 100 kW, the charging time per 100 km of driving distance for heavy-duty vehicles is thus around one hour. According to the EU directive EU 2002/15/EC, the driving time must not exceed 4.5 h, followed by a rest break of 45 min. Under favorable conditions, a truck can drive around 400 km on the autobahn in the driving time. Charging devices for electric trucks would therefore have to be designed for 500 kW, in order to be able to charge the accumulator during the rest break. Vehicles with fuel cells and electric drive have due to their short refueling times operational advantages over pure electric drive, although the overall efficiency is lower than with pure electric drive, since the electrolysis has an efficiency of about 60% and the fuel cell about 50%. Together this amounts to only 30%.

As Table 10.4 shows, trucks and tractor-trailers have together a total share of only 4.1% on the total energy demand of the vehicle sector. Thus, hydrogen electrolysis for heavy-duty traffic is feasible from the perspective of the limited renewable potentials.

Table 10.4 Average energy demand of different drive technologies (power-to-wheel)

Vehicle	Energy share of fuels in 2015	Internal combustion engine 2015 ICE	Accumulator electric vehicle AEV	Fuel cell electric vehicle FCEV
		kWh/100 km	kWh/100 km	kWh/100 km
Car	64.6%	68	16	50
Light duty vehicle < 3.5 t	11.3%	136	28	95
Truck	1.83%	398	80	260
Semi-trailer	2.27%	372	75	250

10.5   Potentials of Sector Coupling

The sector coupling is determined by the potential of renewable energy available in the future. Table 10.5 shows the percentage shares of individual sectors in the total final energy use by application type in Germany [BMWi 2017]. The final energy demand in 2015 was 2466 TWh/a. The share of electricity was 21%.

The heating sector with the subsectors space heating, hot water, process heating, air conditioning and process cooling has a share of 55.85% of the total final energy demand.

The space heating demand can be reduced by 80% through thermal insulation of buildings and by switching to heat pumps, electrical heating and bivalent solar thermal systems. Industrial process heat consists of 85% high-temperature heat above 300 °C (Fig. 10.9) and 88% is generated using fossil fuels and only 12% using electricity.

The renewable generation of high-temperature heat with sufficient potential is currently not guaranteed.

Mechanical energy is the second largest type of application with 38.62%. While industry and tertiary sectors have largely electrified mechanical drives, in the transport sector electricity only has a share of 2% through railways, 4% is accounted for by biofuels and 94% is provided from fuels made from mineral oil. There is great potential here for sector coupling for the electric propulsion, which can reduce the energy demand in the transport sector by 70%. In 2016 electricity generation had a share of 18 to 25% of the end use in most industrialized countries. Figure 10.10 shows an average value of 20%. As the figure shows, the demand in each of the sectors of mechanical energy, space heating and process heat exceeds the value of the electricity generation. Taking into account economic affordability and environmental compatibility, the potential of renewable electricity generation is about twice today's electricity demand, i.e., 40% of today's final energy demand.

Table 10.5 Final energy consumption in Germany by sector and type of application [BMWi 2017]

2015	Industry	Business, trade services (BTS)	Household	Traffic	Total
Space heating	1.96%	7.31%	17.62%	0.14%	27.03%
Hot water	0.18%	0.72%	3.71%	0.36%	4.97%
Process heat	18.89%	1.04%	1.58%	0.00%	21.52%
Air conditioning & process cooling	0.38%	0.66%	1.24%	0.05%	2.33%
Mechanical energy	6.70%	2.75%	0.19%	28.98%	38.62%
ICT	0.37%	0.99%	0.91%	0.11%	2.38%
Lighting	0.42%	2.17%	0.45%	0.12%	3.15%
Total	28.9%	15.6%	25.7%	29.8%	100.00%

Fig. 10.9 Industrial process
heat demand [Bradke 2009]

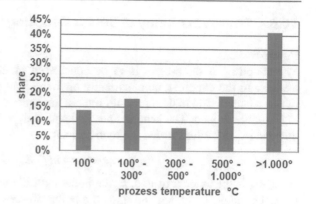

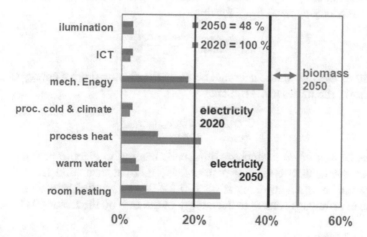

Fig. 10.10 Renewable generation potential and efficiency potential up to 2050 as a percentage
of the final energy demand by 2020

Renewable electrification is therefore not possible without a significant efficiency improvement in the individual types of application. Figure 10.10 shows an efficiency scenario for the year 2050, in which the mechanical energy of the transport sector is largely reduced to 47% of its current demand by switching to electrical drives and intermodal transport concepts.

In the space heating application, a reduction to 27% of today's demand is possible through thermal insulation and heat pumps. With biomass for heating purposes and as biofuel, a share of 8% to 48% of today's final energy demand is possible. Thus, a full renewable supply is theoretically possible. As has already been shown, a full renewable supply of up to around 85% is technically possible, since long-term storage is not feasible either in terms of potential or economic viability. With this efficiency scenario, a sector coupling of power-to-mobility and power-to-heat is possible.

10.6   Energy Economy of Sector Coupling

Power-to-heat

The coupling of the sectors is to be considered in the area of low-temperature heating. In the coupling, an electrically operated heat pump is to be used instead of a gas heater. The building should remain unchanged in terms of its annual heating energy demand. The heating energy demand E_H results from the specific heat demand w_H in kWh/m²/a and the living area A_H.

$$E_H = w_H \cdot A_H = m_B \cdot h_o \cdot \eta_H \tag{10.1}$$

Here, m_B is the amount of fuel, h_o the upper calorific value, because it is supposed to be a condensing boiler, and η_H is the boiler efficiency. If this heating system is replaced by a heat pump with the coefficient of performance ε_{wp}, this results in an annual electricity demand $E_{el,wp}$ for this.

$$E_{el,wp} = \frac{w_H \cdot A_H}{\varepsilon_{wp}} \tag{10.2}$$

The ratio of the substituted heating energy to the electrical drive energy of the heat pump shows the increase in efficiency.

$$\frac{m_B \cdot h_o}{E_{el,wp}} = \frac{\varepsilon_{wp}}{\eta_H} = \frac{4}{0.8} = 5 \tag{10.3}$$

This means that when switching from fossil heating to electric heat pumps, only a fifth of the heating energy is required. By thermal insulation from an average initial value of the heat demand of 160 kWh/m²/a to 20 kWh/m²/a, an overall increase in efficiency related to the energy of the fuel by the factor 40 is possible.

Power-to-Mobility

Electric passenger cars have an average energy demand of 15 kWh/100 km. In comparison, vehicles with gasoline or diesel engines have around 50 kWh/100 km. By switching to an electric drive, the energy demand is reduced to 30%.

In long-distance transport, vehicles with fuel cells and hydrogen tanks will be of interest in the future because of the short refueling times. Hydrogen electrolysis including storage has an average efficiency, based on renewable electricity, of 50%, together with the efficiency of the fuel cell, the overall efficiency is only 25%. Taking into account the charging efficiency of purely electric drives, the efficiency relative to the electric car is only around 30%.

Power-to-Gas-to-Power

The production of synthetic methane from renewable electricity is possible with an efficiency of around 35% to 45%. The reconversion of electricity via a gas and steam power plant or a high-temperature fuel cell with efficiencies of 60% results in overall efficiencies of 21% to 27%. The advantage of this technology lies in the possibility of storing synthetic methane in the pore storage of exploited natural gas deposits. The disadvantage is the low efficiency compared to direct use, for example in the heating sector.

From an energy perspective, there are costs for four successive process steps consisting of hydrogen electrolysis, methanation, storage and reconversion.

10.7 Summary

The coupling of the electricity, heat and gas sectors enables high efficiency levels in the sustainable energy supply. Renewable energy generation is associated with high output power, which at times can exceed the peak load of the grids to above several times. The volatility of the generation characteristic requires flexible behavior in energy use.

The potential for flexible load behavior can be increased significantly through sector coupling. This can increase the usage rate of surplus energy.

Power-to-Heat represents a significant potential. Direct heating, heat pumps and electrode boilers can be used to generate heating and process heat. Electric heating systems are highly efficient and can be used without delay.

In the case of decentralized house heating systems, it is necessary to control groups of consumers in order to be able to adjust the heating load in each case to the surplus production. Large-capacity electrode boilers as supplements in combined heat and power or in plants for industrial process heat can flexibly adjust their output and vary it rapidly over a wide range of settings. They can be connected directly to the medium-voltage grid and have low investments.

Power-to-Gas enables hydrogen to be produced by electrolysis using electricity. In long-distance transport with fuel cell vehicles, this enables rapid refueling and avoids very high electric power fast-charging stations. Synthetic methane (SNG) can be produced through catalytic methanation. This enables long-term, inexpensive storage of large quantities in exploited natural gas reservoirs. SNG can be used in thermal combined cycle power plants with a high degree of efficiency to provide balancing energy with CHP in the event of a lack of renewable generation, or in buildings with gas heating or micro-CHP fuel cell systems.

Power-to-Mobility uses the electric vehicle, which can be charged flexibly according to the availability of renewable electricity. The switch from the combustion engine to the electric drive reduces the energy demand in the mobility sector by around 70%. This makes it possible to reduce the demand in this sector as well, which enables a predominantly renewable energy supply that is tailored to the low renewable potential.

References

[BMWi 2017]   Energieeffizienz in Zahlen. Bundesministerium für Wirtschaft und Energie (BMWi), Berlin (Mai 2017)

[Bradke 2009]   Bradke, H.: Energieeffizienztechnologien in der Industrie, Gewerbe, Handel und Dienstleistungen. Fachkonferenz Energietechnologien 2050, Berlin (2009)

[DESTATIS 2017]   Transportleistungen und Energieverbrauch im Straßenverkehr 2005– 2015. Umweltökonomische Gesamtrechnung. Statistisches Bundesamt Deutschland (2017).

[Elcore]   Strom und Wärme selbst erzeugen. Die Brennstoffzelle mit höchster Effizienz. Prospekt EL192017, Fa. Elcore GmbH, München

[enpros 2014]   Anderloh, T., Graßmann, A.: Flexibilisierung der Betriebsweise von Heizkraftwerken durch Wärmespeicher und Elektrokessel. Kraftwerkstechnisches Kolloquium Dresden (2014)

[FfE 2017]   Flexibilisierung der Kraft-Wärme-Kopplung. Forschungsstelle für Energiewirtschaft (2017)

[Parat]   Hochspannungs-Elektrodenkessel. Technische Spezifikationen, Fa. Parat.

[PFI 2013]   Gewinnung von Prozesswärme aus Abwärme—W2PHeat. AiF Forschungprojekt CORNET 63 EN

[VITOVALOR]   VITOVALOR 300-P, Mikro-KWK auf Brennstoffzellenbasis mit integriertem Gas-Brennwertgerät. Datenblatt der Fa. Viessmann.

[ZSW 2017]   Brinner, A.; Schmidt, M.; Schwarz, S.; Wagener, L.; Zuberbühler, U.: Technologiebericht 4.1 Power-to-gas (Wasserstoff). In: Wuppertal Institut, ISI, IZES (Hrsg.): Technologien für die Energiewende. Teilbericht 2 an das Bundesministerium für Wirtschaft und Energie (BMWi). Wuppertal, Karlsruhe, Saarbrücken (2017)

[ZSW 2018]   Schmidt, M.; Schwarz, S.; Stürmer, B.; Wagener, L.; Zuberbühler, U.: Technologiebericht 4.2a Power-to-gas (Methanisierung chemisch-katalytisch). In: Wuppertal Institut, ISI, IZES (Hrsg.): Technologien für die Energiewende. Teilbericht 2 an das Bundesministerium für Wirtschaft und Energie (BMWi). Wuppertal, Karlsruhe, Saarbrücken (2018)

Options for Action

11

11.1 Options for Action for Building Efficiency

Efficient buildings with low heating energy demand already represent development goals in the EU. In addition to low heating energy demand, buildings should be designed for sufficiency of use with adequate living space per resident. The need for extensive expansion of the photovoltaics integrated into the building will affect most buildings in the future. Solar architecture should be used to develop building forms that allow a high degree of utilization of PV on roof surfaces and facades. Solar thermal energy with vacuum tube or flat-plate collectors should also become architectural attributes on buildings in the future. Roof surfaces should emphasize southern, eastern and western orientations and consider shadow effects from dormers, tall trees or neighboring buildings. Solar architecture has different manifestations in urban high-rise or rural low-rise buildings. Windows in stairwells should be designed to accommodate transparent solar modules.

Efficient heating systems with heat pumps, which in bivalent design are also suitable for cooling in summer, should become the standard. Storage heaters can also be installed in thermally highly insulated buildings. They should be designed as weekly storage systems and avoid overheating of rooms due to the heat loss by high thermal insulation standards. By increased thermal storage capacities, coupling of the sectors of electricity and heat is possible, and surplus energy can be used with high efficiencies and reduce or even avoid the need for fossil fuels.

Through efficient buildings with low heating energy demand and with insulation and efficient heating systems, it is possible to reduce the energy demand of this sector, which enables to get along with the low renewable generation potentials.

© Springer Fachmedien Wiesbaden GmbH, part of Springer Nature 2022
G. Brauner, *System Efficiency by Renewable Electricity*,
https://doi.org/10.1007/978-3-658-35138-0_11

11.2   Options for Action for Efficient Mobility

The energy demand of the transport sector with vehicles with combustion engines roughly corresponds to the regenerative potential in almost all European countries. The switch to electric vehicles in suburban and rural local transport can lead to a reduction of the sectoral energy demand by 70%. At the same time, the vehicles can be supplied with renewable electricity from wind energy, photovoltaics, hydropower and biomass. By developing automated vehicles and digitization, a flexible car-sharing can be created that enables mobility for everyone, even without driving license. The small automated electric vehicle with digital user communication can revolutionize suburban and rural local traffic.

In heavy-duty long-distance transport, the purely electric drive still has a need for development because of the high charging powers around 500 kW required. Heavy-duty vehicles with electric drive and fuel cells can be operated with electrolysis hydrogen, produced from renewable surplus electricity. User acceptance is expected to be high due to the short refueling times. Heavy-duty vehicles with electric drive and supply via overhead traction line on designated routes are also possible.

In urban areas, local public transport is the most efficient form of mobility. In long-distance transport, emission-free and regenerative mobility is possible with electric trains and electric buses.

11.3   Options for Action in Industry and Commerce

In the industrial and commercial sector, the drives for mechanical energy are already largely equipped with electric motors and are therefore compatible with the future renewable generation structure. In the area of process heat, fossil fuels are still predominantly used. Here, electric boilers and electrode boilers can create a sector coupling for the use of surplus electricity for the generation of low-temperature process heat and thus substitute fossil fuels. Biomass has significant potential, which is in the order of magnitude of renewable electricity. By converting biomass into process gases such as hydrogen, biomethane or liquid biofuels, it is possible to supply industry with sustainable fuels for high-temperature heat supply.

11.4   Options for Action in the Household

Households have no direct targets for efficiency. Efficiency is to be achieved indirectly. Energy suppliers are required to provide top-down incentives to their end users to reduce their energy demand through efficiency measures. The European Ecodesign Directive has a bottom-up effect. It prescribes efficiency standards for appliances that may be manufactured and traded in Europe. In the longer term, this will ensure that only appliances with minimum efficiency standards are used. Efficiency in the household enables an affordable and sustainable energy supply. This means that in the future energy expenditure can remain at a level similar to that before the energy transition. At the same time, this ensures that the existing renewable energy potentials are sufficient for a full supply due to the reduced energy demand.

11.5   Options for Action for Regenerative Generation

11.5.1 Wind Energy

Wind turbines can be installed in the sea (offshore) or on land (onshore). On land, low-wind turbines should be increasingly used in regions remote from the coast. They require reduced grid expansion due to smaller generator power with higher full load hours. Because of the larger rotors and thus reinforced towers and foundations, they have higher investment costs. However, the sum of investment from low-wind energy and grid expansions remains roughly unchanged. Reduced grid expansion also finds higher acceptance. In the future, wind energy will be the most important regenerative generation technology in Europe with the highest growth rates. The most favorable mix of renewable generation consists of 75% wind energy and 25% photovoltaics. This enables a coverage rate of 80% of the households loads if they have local storage for photovoltaics.

11.5.2 Photovoltaics

In future. photovoltaic is the second most important generation technology after wind energy in Europe. For reasons of environmental protection, in the future PV systems will be installed mainly on roofs and facades of buildings. The PV power should be tailored to the needs of the residents. On the one hand, this will prevent the distribution grid from being overloaded. On the other hand, it is expected that photovoltaics will no longer be subsidized by feed-in tariffs in the future. The profitability of the systems will be based on the avoided electricity purchase costs from the grid. The contracting of PV systems is an interesting future measure to give everyone access to sustainable energy supply at reduced electricity tariffs.

11.5.3 Hydropower and Pumped Storage

From a supply perspective, hydropower is the most favorable form of renewable generation. It has a guaranteed minimum power of about 40% and a more uniform supply without short-term fluctuations. However, its remaining potential is limited. Potential reductions are to be expected from the European Water Framework Directive.

Pump storage is gaining in importance. They are needed to absorb surplus energy and can feed it back as balancing energy in the event of a shortage of renewable generation. In addition to thermal power plants, they are important for security of supply in the event of so-called dark doldrums. Pumped storage facilities should be designed in terms of its capacity as weekly storage facilities. In the past, they were only designed as day storage for hydrothermal combined operation. The new design is necessary because wind energy in particular can cause periods of high generation over several days. Larger rolling volumes also improve the flexibility of the pumped storages, since they can then be used for several days in the direction of pump or turbine operation even when the filling level is half.

11.5.4 Biomass

Biomass has great potential that can be used directly, e.g., for heating. In addition, there are options for the production of biogas or liquid fuels. This increases the renewable potential available and reduces the pressure to increase efficiency in the electricity application sector.

11.6   Options for action in Energy Grids

The energy grids have the task of connecting large capacities to regenerative energy sources. They take over the transport from the regions with high generation to the load centers. Since the renewable energy sources have high power with lower values of annual energy, the transmission capacities of the energy grids must be at least doubled. Since no higher transmission voltages than 380 kV can be built in densely populated Europe, the strategy for grid expansion consists is to reinforce existing grids and build selected new DC links over longer distances. As reinforcement, the conversion of existing 220 kV overhead lines to 380 kV is possible. This can increase transmission capacity at minimum by the factor of three using the same route with the same mast height. Overhead lines also have the advantage that, in the event of high wind generation, the higher wind speeds improve the cooling of the overhead line conductors, thereby tripling the capacity. Cables do not show these properties. They are not technically suitable for the 380 kV transmission networks. In the medium- and low-voltage grids, on the other hand, cables can be used very well.

11.7   Human Factors

From the perspective of the available technologies, a predominantly regenerative energy supply is possible. This requires efficient and sufficiency-oriented behavior in the end use of energy. Furthermore, broad acceptance is required for the expansion of infrastructures in the area of grids, pumped storage, wind turbines and building-integrated photovoltaics. Efficient end-use and efficient, i.e., cost-effective, infrastructures enable a renewable energy supply that makes do with the existing potentials and is comparable in terms of costs with those before the energy transition. A broad awareness is necessary to make this possible. This book is intended to help this.

Printed in the United States
by Baker & Taylor Publisher Services